CARTOGRAPHY
IN FRANCE
1660–1848

CARTOGRAPHY IN FRANCE 1660–1848

SCIENCE, ENGINEERING, AND STATECRAFT

Josef W. Konvitz

With a Foreword by
Emmanuel Le Roy Ladurie

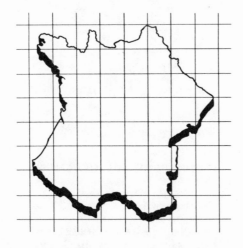

The University of Chicago Press · Chicago & London

Josef Konvitz is professor of history at Michigan State University. *Cartography in France, 1660–1848: Science, Engineering, and Statecraft* received the Nebenzahl Prize in the history of cartography in 1985.

The University of Chicago Press, Chicago 60637
The University of Chicago Press, Ltd., London

© 1987 by The University of Chicago
All Rights reserved. Published 1987
Printed in the United States of America

96 95 94 93 92 91 90 89 88 87   5 4 3 2 1

Library of Congress Cataloging in Publication Data
Konvitz, Josef W.
    Cartography in France, 1660–1848.
    Bibliography: p.
    Includes index.
    1. Cartography—France—History.   I. Title.
GA861.K66   1987   526'.0944   86–11283
ISBN 0–226–45094–5

For Eli and Ezra

Live with your century, but do not be its creature; render
to your contemporaries what they need, not what they
praise.

Friedrich von Schiller,
On the Aesthetic Education of Man in a Series of Letters

Contents

Illustrations

Full citations appear with the legends in the text. All maps whose source is given as Bibliothèque Nationale, Paris, are from the collection of the Département des Cartes et Plans unless another collection is listed.

Foreword

Emmanuel Le Roy Ladurie

In an important thesis on D'Alembert and Diderot, Jacques Proust spoke fondly of "manufacture encyclopédique." Shall we say, in the same spirit, that "manufacture cartographique," of which Josef Konvitz, excellent authority on the French Old Regime, has become the apologist, was in the beginning a creation of Colbert? At any rate, let us note the great contrast, from a cartographic point of view, between the underdevelopment of France during the reigns of Henry IV and Louis XIII and the enormous progress the country made in this field after the personal accession of the Sun King and the installation of his hardworking minister Jean-Baptiste Colbert in the corridors of power. Let us not, however, sing the solo of national glorification! The Italian and British contribution during the Renaissance and the baroque era was considerable. Without their clarification of triangulations and logarithms, the new cartographic discipline would not have gotten off the ground.

Colbert's impetus gave geographers and astronomers much to do. From the seventeenth to the nineteenth century, the name and lineage of the Cassinis symbolized this double undertaking, with geocosmographic claims. Realistic understanding of space, at the time, was accompanied by the perfection of terrestrial concepts. The flattening of the earth at the poles was demonstrated at great cost, and the French realm was covered, on the ground and on paper, with hundreds of theoretical triangles. Peasants were sometimes alarmed at the sight of geodetic teams laden with strange instruments. After the witches!

A few bureaucratic or official figures (Orry, Trudaine) and the remarkable administrative and educational institution of the Ponts et Chaussées guarded the achievements of cartographers in the time of Louis XV, a king too long misunderstood. The two generations of "Cassini" maps, one finished about 1745, the other gloriously completed during the last decades of the Old Regime, still testify in libraries today to the power of an accomplishment as much intellectual as empirical. At the same time an international collaboration asserted itself by design—not, as in the previous century, by chance. During the reign of Louis XVI, the French and English cooperated in extending vast triangles above the watery expanse of the Straits of Dover. Was it chance that the Treaty of Commerce of 1786, easing restrictions between the two countries, was concluded in the same period? On a theoretical level, in the field of mapmaking France maintained a slight lead over Great Britain, but the latter was already producing outstanding graphic specialists; the French government offered some of them great sums of money to leave their country.

Beyond chronological considerations, Konvitz tackles ideas concerning international boundaries. In the sixteenth century the boundary was only a vague, blurred swath. It was narrowed into a precise dotted line running along the rivers when the treaties concluded by Louis XIV were finally drafted accurately and when an office of boundary demarcation was established in the ministerial headquarters. One saw great minds at work in the course of establishing borders—the minister Vergennes, for example; one also saw swindlers bustling about, some competent enough, like the picturesque Rizzi-Zannoni, whom Konvitz rescues from two centuries of obscurity. Cartographic "lobbies" emerged; some, which worked for the army, quarreled with specialists from the departments of Foreign Affairs and Public Works.

The progress of French cartography at the time of the Enlightenment was linked to collaborations between state and science. They were embodied in other areas (medicine, chemistry, exact sciences, statistics) by Vicq d'Azir, Condorcet, Lavoisier.

Taxation (and especially property taxes, which hit agriculture ipso facto) also had its say, as in the conspicuous example of cadaster making. This time the great man was not Colbert but Turgot, an economist of considerable merit. Intendant of Limousin, he introduced provincial cadastration. Similarly, let me point out Bertin, whose influence was so great in deregulating the grain trade. And let me finally mention Babeuf, who compiled seignorial or fiscal atlases. He later became the inventor of a communist ideology that would lead him to the guillotine. The cadastral undertaking still interfered with other decisions. It cut the territory of the country into ninety departments and, moreover, imposed the revolution of weights and measures, objectified in the standard meter. Savoy (which was not yet French) had demonstrated, well before 1789, the efficiency of an avant-garde cadaster. Just annexed, Corsica was one of the first zones mapped, in spite of an insular type of underdevelopment. One finds yet today in the dusty archives of Continental city halls a cadaster that is readily called Napoleonic. In reality, its preparation was pursued during the entire first half of the nineteenth century. This document marked the definitive beginning, by the postconsular bureaucracies, of projects of fiscal equity initially conceived by Turgot and Bertin. Once begun, it took half a century to finish the first edition of French army maps to a scale of 1:80,000. Their hachures have fascinated geography students until modern times.

One of the most original aspects of Konvitz's contribution concerns the mapping of seas and ocean floors. Once again the personality of Colbert was fundamental, for it again signaled the orientation toward the scientific, academic, and centralized state that asserted itself in many other sectors of police, finance, and so forth. The construction of a national war fleet was an expanding industry for 130 years before the Revolution. Draftsmen of sea charts had their special place in this extensive work. Instead of hachures, they used the bold technique of contour lines, which would take two centuries to be adopted in all areas of terrestrial as well as oceanic cartography. Social history did not remain a stranger to this development of knowledge. The creation of the Naval Academy at Brest, after 1751, established the eminence of a scientific nobility including officers of the Royal Navy. As a group, they constituted one segment of the society of the Enlightenment. The English, a practical people, kept the upper hand with a superior fleet. But the French, Cartesians as always, surpassed the British when it came to making naval maps.

An excursus into mountain cartography gives Konvitz the opportunity to retrace the birth of a rigorous geology and mineralogy. This time the forefather (surprise!) was no longer Colbert but Vauban, whose fortresses dotted the plains and the steep passes of the mountainous zones. The famous relief maps, whose removal in 1986 provoked some controversy, are one of the remainders (among others) of this new approach of a mountain strategy. For once the great Saussure, explorer of the Alps, did not distinguish himself much in this field. A good writer, tireless traveler, and excellent narrator, he was not a true cartographer. In the astute network of the Fortifiers of France, the harbingers of pure science and of applied knowledge occupied a fundamental position. As far as they are concerned, let it suffice to mention the birth of descriptive geometry, under the auspices of Monge, and the efflorescence of the powerful laboratory of practice and ideas that was the School of Engineers of Mézières. Konvitz is far from walking in the footsteps of today's philosophers who, seeing repression everywhere, reduce the admirable pedagogy of the Grandes Ecoles of the Old Regime to acts of oppression and the reforming of youth.

Now on to maps of canals. The Colbertian Riquet, "inventor" of the canal between two seas, unintentionally created a need to represent waterways. The mapping of canals developed after his death. In spite of the innovations of which Riquet was the precursor, France would later have few canals but many maps of their courses. Less-cerebral England would occupy itself with digging rather than drawing them.

As for roads, the effort of government agents of Versailles was clearly more balanced. The monarchy cleared postal routes that served the army at the frontiers and gave a boost to commerce. At the same time the technocrats of authority noted on paper the road configuration. Road mapping would later reach its apogee thanks to the Michelin map, a classic of its kind in the world.

Finally, agriculture, industry, meteorology, and statistics carry the inventory of this book to an iconology of space, a domain in which a Dupin and a D'Angeville had distinguished themselves as early as the nineteenth century. Long misunderstood, these two men were the cartographers of both regional underdevelopment and development. Josef Konvitz thus continues and even broadens the pioneer work of François de Dainville. From the time of this Jesuit discoverer of cartographers of the past, to the young researcher from Michigan State University, the reader passes through a well-documented era, thanks to investigations that happily marry precision with erudition.

Preface

My interest in the history of cartography grew out of my study of port city planning in early modern Europe, the subject of my dissertation and my first book. In the course of examining hundreds of maps and plans of cities from the Baltic to the Mediterranean, I noticed a dramatic improvement in cartography in France between the 1660s and the 1690s. At first I did not attempt to explain how or why this change occurred, since such questions were peripheral to my main topic. Then the Newberry Library granted me a one-month fellowship in 1976 to study how statesmen, merchants, and navigators used maps and atlases in the seventeenth century. While engaged in this research, I observed that the most recent monograph on cartography in eighteenth-century France had been published in 1905.[1] Encouraged by John Long, Adele Hast, Robert Karrow, and David Woodward, I decided to study the development of cartography in France from the 1660s to the 1840s. A sabbatical leave from Michigan State University and a Fellowship for Independent Study from the National Endowment for the Humanities in 1979–80 provided indispensable support for a year of research in France. A grant-in-aid from the American Council of Learned Societies sustained an additional three months of study in 1982 in Belgium, Scotland, and France.

Librarians and archivists in numerous institutions aided my research. Their help was all the more necessary because many of the documents I wanted to see had not been consulted for decades and required delicate handling. Thanks to Monique Pelletier and her staff, I was able to organize my research around the vast collection of the Département des Cartes et Plans of the Bibliothèque Nationale. Its superb reading room is a splendid setting for work. The personnel of the Cabinet des Estampes, the Cabinet des Manuscrits, the Salle des Imprimés, and the Service Photographique of the Bibliothèque Nationale provided invaluable assistance. I also wish to thank Mme Falkay and her staff of the map division of the Archives Nationales. Many maps in the Archives Nationales, however, are in cartons of written documents stored outside the map division. These cannot be examined efficiently and carefully because desks of the main reading room are too small and too poorly lit and because the policy on the number of documents that may be consulted at one time is too restrictive for research in the history of cartography. Given the amount of material I had to study elsewhere, I decided to make only a cursory inspection in the Archives Nationales of the maps and papers of the French navy's hydrographic office and of maps attached to documents from the Ministry of the Interior. I believe that research in these materials would deepen our knowledge of institutional practices and thematic trends but would not modify the main outlines of the story presented in this book. Elsewhere in Paris I was aided by Marc Duranthon and his colleagues at the Institut Géographique National; by Jean Michel and his associates at the Centre Pédagogique de Documentation et de Communication of the Ecole Nationale des Ponts et Chaussées; by the staffs of the Archives de l'Académie des Sciences, the Bibliothèque de l'Institut, the Ecole Polytechnique, the Ministère des Affaires Etrangères, and the Observatoire de Paris; by Nelly Lacrocq and her staff at the Archives de Génie; and by their colleagues of the Service Historique de l'Armée de Terre and of the Service Historique de la Marine. In Scotland I received assistance from Mr. Richard Dell and the staff of the Strathclyde Regional Archives and from the staffs of the Mitchell Library, the Scottish National Library, and the Scottish Record Office. In Belgium I was hospitably received at the Académie Royale des Sciences.

Many friends have contributed ideas, asked useful questions, and sustained my enthusiasm. I record with pleasure my debts to the late David C. Bailey, Marc and Elizabeth Castellazzi, Steven and Molly Cobb, Roy and Toni Flemming, Larry and Helen Foster, Bruno Fortier, Christopher and Rhoda Friedrichs, Steven Kramer, Emmanuel and Madeleine Le Roy Ladurie, Harold Marcus, Michel and Mireille Pastoureau, Mary Pedley, Theodore K. Rabb, David and Rudite Robinson, Les and Kathleen Rout, L. Pearce Williams, and John and Susan Woodbridge. David Buisseret, Director of the Hermon Dunlap Smith Center for the History of Cartography at the Newberry Library, and three anonymous readers provided valuable suggestions for improvement. My wife photographed many maps and helped me find the time to write and revise successive drafts of the manuscript on my word processor. We gratefully acknowledge the anonymous designers at Leica and Digital whose skills have enabled us to work more productively. Le Roy Ladurie's Foreword was translated by Gail Riley.

Portions of two earlier articles of mine have been included in this book. "The National Map Survey in Eighteenth-Century France" is reprinted with permission from *Government Publications Review* 10 (1983): 395–403 (copyright 1983, Pergamon Press, Ltd.); "Redating and Rethinking the Cassini Geodetic Surveys of France" is reprinted with permission from *Cartographica* 19 (1982): 1–15 (University of Toronto Press).

Introduction

This book focuses on three interrelated aspects of cartography in France between 1660 and 1848: the national map survey, maps of landforms and hydrographic mapping, and thematic cartography. Scientists appear prominently in the discussion of the national map survey; the development of new techniques for maps of landforms and hydrographic mapping, and especially of the contour line, highlights differences in the approaches of scientists and engineers; thematic mapping was largely the achievement of engineers. The interaction between cartography and public affairs is the theme that holds the book together, relating normative and innovative practices and the diffusion of new techniques to cultural, technical, and political factors external to cartography.

The French state was not a monolithic entity but a constellation of evolving institutions and functions. The process of political change often concentrated upon the relation between centralized and decentralized authority. In the context of this study, the state refers to the government of France, not to provincial and local authorities. In 1961 Rhoda Rapoport considered "the question of governmental attitudes toward science" to be "perhaps the greatest remaining gap in our knowledge" about science in eighteenth-century France.[1] Charles Gillispie's *Science and Polity in France at the End of the Old Regime* is a major contribution toward closing that gap,[2] but because Gillispie covered many subjects, he was obliged to rely upon a few secondary sources for information about cartography and to avoid references to the more specialized kinds of mapping that represent the major portion of this book. Gillispie would be the first to recognize that the achievements of French cartographers merit further study.

The history of cartography is a young academic subject whose methods and objectives are still in a formative stage.[3] As a historian, I have been more comfortable with the study of the political and institutional aspect of French cartography than with the technical process of mapmaking. I hope that other scholars will study and write about some of the topics I have not covered. Most people already familiar with eighteenth-century French maps have seen atlases or maps edited by Delisle, Jaillot, de Beaurain, d'Anville, Le Rouge, Vaugondy, and their families and associates.[4] Such maps and atlases will not receive attention here because they did not contribute to any of the three aspects of cartography featured in this book. The French map trade began to decline on the eve of the Revolution, but some of the most interesting developments in French cartography initiated during the Enlightenment had their greatest impact only after 1789. Although the methods of the great map editors and printers were unable to meet the needs of the state for new kinds of maps of unprecedented coverage, their efforts to synthesize geographic knowledge and graphic art remain impressive on their own terms and merit further investigation. An essay I am writing with Mary Pedley for the multivolume *History of Cartography* being published by the University of Chicago Press will attempt a synthesis of knowledge about public-sector cartography and the private map trade. Several other topics, many of which could be the subjects of monographs, have been omitted as not relevant: estate surveying, celestial cartography, maps as reconstructions of historical events, Napoleonic mapping, French maps of other countries, maps in the context of topographical literature and illustration, and the technical process of designing and printing maps. Another topic that merits further study is the production and use of military maps for supply and transport, strategic planning, and battlefield control. At the risk of leaving out some topics that describe routine cartographic practices and some publishing efforts that interest collectors, I have concentrated upon how and why new ways of making and using maps were introduced in France between 1660 and 1848.

French scientists, engineers, and public officials were responsible for the most important and distinctive innovations in cartography in eighteenth-century Europe. They dominated cartography at that time as no other national group has done since. People in other nations imitated and improved upon their methods, so that qualitative differences between maps made in France and in other countries diminished during the first half of the nineteenth century. Yet even then the French continued to make significant contributions, especially in thematic mapping. By expanding the analytical uses of maps, by establishing unprecedented standards of accuracy, and by nurturing institutional frameworks to sustain mapping projects over many years, the French contributed to one of the central concepts of modern times: that through direct observation and accumulated information man can better understand and manage his affairs.

A Note on Terms

Throughout the period covered by this book, the term cartography did not yet exist. People spoke of maps as *cartes* or *cartes géographiques*; they also referred specifically to plans, surveys, marine charts, topographical drawings, and eventually to thematic maps (*cartes figuratives*). Many mapmakers who considered the problems of design and intelligibility conceptualized how maps communicate information and how maps are used, but no one apparently felt the need for a term that embraced the diversity of maps and of the specialized activities related to their preparation and use. Writing in Paris in 1839, Manuel Francisco de Barros y Souza, second viscount of Santarem, coined the word *cartographie* and used it in a letter. The word was printed for the first time in the *Bulletin* of the Société de Géographie de Paris in 1840; "cartography" appeared in print in English in 1843.[1] For convenience, I use the words mapmaker, cartographer, and cartography in their modern, colloquial sense. Mapping can be understood today to produce a mental image of space, and it is frequently used as a metaphor for the acquisition of knowledge,[2] but I use the term in a more restrictive sense to mean the production of a graphic, tangible image.

Map scale is another cartographic concept I have converted from eighteenth-century terms into modern units. A large-scale map shows a small area in great detail, and a small-scale map shows a large area with far less detail. In the eighteenth century scale was described as so many *pouces* or *lignes* on the map equivalent to so many *toises* on the ground. (One toise was approximately equivalent to 2 meters, and one league to 4,444 meters.) Ever since the Napoleonic revision of cartographic standards in 1802–3, the equivalence between the scale of a map and a unit of measurement on the ground has been called the representative fraction. In this book the representative fraction is a multiple of the centimeter. Thus, 1:10,000 means that each centimeter on the map is equivalent to 10,000 centimeters on the ground. Here are several standard eighteenth-century scales expressed in metric representative fractions, together with their typical use:[3]

12 pouces	= 100 toises	1:600 cm	Field surveys for relief maps
3 pouces	= 100 toises	1:2,400 cm	Field surveys for constructing fortifications
8 lignes	= 100 toises	1:10,800 cm	Geodetic field surveys
4 lignes	= 100 toises	1:21,600 cm	Military reconaissance
1 ligne	= 100 toises	1:86,400 cm	Cassini map surveys
1 pouce	= 3 leagues	1:493,000 cm	Printed map of a province
1 pouce	= 15 leagues	1:2,465,000 cm	Printed map of a nation

One

The National Map Survey

1660–1730

When Louis XIV came to power following the death of Mazarin in 1661, the future of cartography in France did not look promising. Henry IV and Louis XIII, his grandfather and father, had tried to promote cartography but had little success.[1] The French were still far behind the Dutch in mapmaking and behind both the English and the Dutch in such related fields as navigation, applied mathematics, and instrument making. Yet within a few years, interest in cartography began to wane in England and the Netherlands, only to blossom in France. The development of cartography in France was all the more spectacular because the French did not slavishly imitate the English and the Dutch. Using techniques and principles first made known by others, they transformed the methods of mapmaking and the contexts in which maps were used. The reputation of the French rested on the national map survey, executed under the direction of the Cassini family from 1681 until the Revolution. The first patron of the survey was Colbert.

In 1661 Jean-Baptiste Colbert (1619–83) was granted wide authority by Louis XIV in naval, economic, and cultural affairs. Colbert knew he could exercise great power in the king's name if he used that power to enhance the interests of France and the stability of the monarchy. Colbert and Louis, eager to restore France's economic and political fortunes after wars with Spain and a long period of domestic insurrection had weakened the authority of the state and depressed economic activity, set about reforming French institutions and attitudes. One of the first steps Colbert took was to order an inventory of France's resources. This brought him to realize what he might well have suspected but must have hoped was not true, that French maps were deficient in both quantity and quality.

In 1663 Colbert instructed officials in the provinces to survey the condition of France.[2] He mentioned maps at the beginning of his instructions, a very lengthy memorandum that touched upon every kind of administrative unit and government function. He asked that the crown's agents send the capital accurate and detailed maps of each province and généralité and have new maps made by capable, intelligent, and skillful persons of areas not covered satisfactorily by existing maps. He further specified that written reports be submitted to describe areas that no map, new or old, covered and that maps and reports alike illustrate the divisions between military, judicial, fiscal, and ecclesiastical administrative units. Colbert wanted reports on everything about the life and institutions of France that could possibly affect the government. Nicolas Sanson (1600–1667), the government's first official cartographer, was to receive all the written and graphic documents submitted in response to these instructions for use in his work.

Sanson's techniques were those of the *géographe du cabinet* active in the print trade; his product was the hand-colored map depicting the state of nations in various epochs since antiquity, to be collected in atlases and consulted at leisure. His maps were based on a critical evaluation of textual sources and comparisons between many maps, but they were expensive to produce and were of too small a scale to be of use in forming and executing public policy. During the first half of the seventeenth century, another way of making maps, based on original field surveys, was introduced in the Netherlands. It generated more detail and reached a higher level of accuracy than had been possible before, but it also absorbed even greater financial and technical resources. Direct observation and measurement of the world, rather than a close reading of scholarly sources, eventually provided cartography with truly scientific standards. Rather than trying to reform Sanson's position and methods, Colbert characteristically retained the old position and created a new institution alongside— the Académie Royale des Sciences (hereafter, the Academy). Although the Academy was not created solely for the purpose of transforming cartography, mapping was one of its first activities, and long one of its most famous.

The first problem the French confronted in mapping their own country was to determine its length, breadth, and shape as accurately as possible, according to the science of geodesy. Gemma Frisius (1508–56), a Dutch physician, provided the first description of geodetic triangulation. Triangulation involves measuring a baseline (one or two kilometers long today) and then constructing a triangle away from the baseline, such that the lengths of the sides can be accurately determined by trigonometry as a function of the angles of the triangle. Using steeples or special scaffolds as landmarks, surveyors might extend networks of triangles one hundred miles from a single baseline. Italian engineers used the principles of triangulation to measure distance and elevation in fortifications in the sixteenth century, and Tycho Brahe used them in 1578–79 to establish the position of Van Island in relation to the rest of Denmark. The first person to complete a major triangulation project was Willebrord Snel van Roijen (1580–1626). The inventor of a quadrant to measure the angular distances of planets and stars, Snel took advantage of the tripod and the circumferator, a surveyor's astrolabe for measuring angles that combined a circular scale with a magnetic compass, both recently invented. Snel described his methods in *Eratosthenes batavus* (1617), but because he still lacked logarithmic tables his calculations were lengthy and tedious. Regiomontanus (1436–76) had composed tables of tangents, secants, and sines to the fourth place; Georg Joachim Rhäticus (1514–76) published tables for all six trigonometric ratios in intervals of ten seconds between zero and ninety degrees to seven places. But their tables were for use only in a scholar's study and would still have challenged even the most determined and capable student of mathematics. Logarithms, originally published by John Napier (1550–1617) and improved by Edward Wright, Edmund Gunter, and Henry Briggs between 1615 and 1628, replaced multiplication and division with addition and subtraction. Napier also popularized the use of decimal fractions and standardized the use of the period in representing decimals. By the end of the 1620s, therefore, the mathematical concepts and tables necessary to perform triangulation routinely were finally available.[3]

Yet geodetic surveys were not quickly launched once better methods became available. Improvements in mathematical computation, such as trigonometric or logarithmic functions to the fifth or tenth decimal place, greatly exceeded the ability of craftsmen to fashion instruments that gave readings of comparable precision. Improvements were needed for direction finding, leveling, measuring angles, and marking circular and linear scales precisely on instruments. In 1571 Leonard Digges made an early theodolite, a composite instrument that determined vertical and horizontal angles, but it did not become important until the middle of the eighteenth century. Similarly, the device Pierre Vernier made in 1631 for dividing or reading linear and circular scale divisions was not applied to surveying instruments for nearly a century, and the spirit level with a bubble, invented as a leveling device by Melchisedech Thevenot in 1666, was neglected for fifty years. William Gascoigne's invention of the micrometer, which permitted the angles between two objects to be read precisely, was forgotten between 1640 and the 1660s. High costs, a small market, and the small number of suitably trained craftsmen retarded the development of better instruments and the acceptance of more rigorous surveying techniques.

The English were at least as well placed as the French to make progress in the practical application of mathematics to cartography, but the indifference of many scientists and inadequate funding compromised their efforts.[4] Despite attention to estate surveys, to better instrumentation, to problems of navigation, and to instruction in mathematics, after the 1620s triangulation was no more in evidence in England than in the Netherlands. Some Englishmen conceived of ambitious projects to map England, but neither the Royal Society (founded in 1662) or the Royal Observatory (founded in 1675) was able to support them. Of the society's early membership, only Edmond Halley (1656–1742) maintained a consistent interest in cartography, but he worked alone and intermittently.[5]

Perhaps a latent interest in mapping was as strong in England as in France, but the right men were not there to exploit it. Later, even when the French made startling advances, the English made no effort to catch up. Yet the English needed better maps as much as the French. Clearly, neither need alone nor better methods were enough to bring about a reform of cartography in England. Informal efforts to improve cartography based upon the occasional collaboration of practical men and men of science were unable to achieve sustained and rapid progress. Given the lack of demonstrable achievement, innovation in cartography no doubt appeared risky. The examples of bold projects that failed would have deterred others. Cartography needed consistent, visionary leadership, the cooperation of many talented and well-trained professionals, and a lot of money. In France, the government under Louis XIV and Colbert began to provide leadership that brought results.

In 1668 Colbert asked the Academy to recommend ways of making more accurate maps of France.[6] Originally the crown intended to house the Academy and its laboratories and collections in the Paris Observatory (erected 1667–72), but in the end the institutions remained physically separate. The role of astronomy in the cartographic work of the Academy was nonetheless very great. The outstanding problem in navigation was finding longitude at sea.[7] To determine longitude, it is necessary to know the time in two places simultaneously, the difference in time being a measure of distance. One solution, to carry a perfect timepiece on board ship, was not available until the middle of the eighteenth century. Another method involved recording the time of some astronomical phenomenon simultaneously in two places and then comparing the results later; alternatively, if the time a phenomenon would be visible in one place was known, once the time it appeared in another place was determined the difference could be calculated immediately. Jean-Dominique Cassini (or Cassini I, to distinguish him from his progeny), was already famous for his work on the longitude problem when Colbert offered him membership in the Academy in 1665.[8] Cassini (1625–1712) had completed tables of the movements of the satellites of Jupiter that made Galileo's earlier work on the subject accessible to navigators. The relation of astronomy to cartography went beyond the longitude issue. As the problem of mapping France was understood in Colbert's day, the size and form of France could be determined only once the size and shape of the earth itself were known.

Cassini arrived in Paris in 1669. In 1671 he moved into an apartment in the observatory, in 1673 he became a French citizen, and in 1674 he married Geneviève de Laistre, daughter of the lieutenant general of the Comté de Clermont, north of Paris, whose dowry included the Chateau de Thury. Cassini had intended his stay in France to be temporary, but he lived there until his death.

In 1668 the scientists of the Academy decided to test several mapping techniques in the environs of Paris under the supervision of Gilles de Riberval and Jean Picard.[9] Picard (1620–82) had experience in making and using instruments for astronomic measurements.[10] He wanted to enlarge the Academy's map work to obtain a more accurate estimate of the earth's circumference by measuring an arc of meridian. Picard was concerned that an error of a few minutes that any ordinary surveying instrument of the period could not detect might have been made at several places, thus throwing off the triangulation over a large area. He wanted each triangle to be verified independently with the best instruments, which he alone had. Picard realized that this work would also permit the measurement of an arc of meridian. Already in 1668, he had experimented with his instruments on a triangle with a base six thousand toises long, and he proposed to measure the distance between Sourdon (near Amiens) and Malvoisine (near Corbeil-Essonnes), thirty-two leagues apart and apparently on the same meridian, by connecting a series of triangles between them. By measuring the latitude of each place, Picard thought he could deduce the length of a line of longitude between them. In 1668 Picard measured one base on the Paris–Fontainebleau highway, between the mill of Villejuif and the pavilion of Juvisy. But he had not measured a base to the north of Paris by the time he met Cassini, and they did so together in July 1669, using a very large instrument (with a ten-foot radius) that had just been assembled. Picard measured thirteen triangles in 1669 and 1670. His base was 5,663 toises in length. He even measured the wooden rods he used to measure the baseline, and he repeated part of his work as a check.[11]

The rapid pace and success of the work undertaken in 1668–70 enlarged Picard's ambition. He proposed to the Academy in 1671 that he travel to Uraniborg to compare the observations Tycho Brahe made there between 1576 and 1597 with astronomical data to be gathered in Paris by Cassini. This involved determining the difference in longitude between the two places using Cassini's tables. During his stay in Denmark (eight months in 1671–72), Picard was assisted by Olaus Roemer, whom he induced to return with him to Paris. Roemer stayed in Paris until 1681, carrying out, among other things, important work concerning the speed of light. Observing the eclipses of Jupiter's satellites, Roemer noticed that when the earth was moving away from Jupiter the time between eclipses was longer than when the earth was moving toward Jupiter. He interpreted this to mean that when the earth is moving away the light emitted at the moment of the eclipse must travel farther and thus take longer to reach the earth. From this Roemer deduced that light travels at a finite speed; his calculations of the speed of light were close to the currently accepted value. This knowledge proved invaluable to Cassini in revising the tables of 1668, as did the work of Jean Richer in Cayenne, French Guiana, in 1672.[12] In the 1693 edition of the tables the time for the eclipses of Jupiter's satellites was given for the Paris Observatory. Thenceforth, longitude could be determined anywhere in the world by comparing the local time for an eclipse with Cassini's tables.

Picard's trip to Denmark and Richer's voyage across the Atlantic were only the first of several expeditions related to cartography undertaken by academicians. Picard traveled to Touraine in 1672 and to Languedoc and Lyon in 1674; Cassini traveled to Provence in 1672. In 1674–75 Picard helped survey the area of Versailles to supply the châteaus of Marly and Versailles with water for fountains and canals, perfecting in the process the application of the bubble level to the telescope. Such activities involving trained scientists represented a "great advance in scientific technique" and in methods of investigation, of significance to the progress not of one branch of knowledge alone, but of science itself.[13] Enough fieldwork had been done to demonstrate that the coordinates of latitude and longitude for many places were erroneously given on most maps.

In 1679 Picard and Colbert obtained Louis XIV's approval for their proposal to make a map of France with the greatest possible accuracy. Picard observed that the coastal fieldwork so critical to hydrographic charts could not be verified and coordinated unless astronomical observations were made by skilled personnel using Cassini's tables. This he proposed to do with Philippe de La Hire (1640–1718), known for his work on conic sections and a member of the Academy since 1678.[14] In 1679, when Picard and La Hire traveled to Nantes and Brest, they found that the position of Brest, France's principal naval base, was thirty leagues too far to the west on existing maps. In 1680 they went to Bayonne, Bordeaux, and Royan (La Rochelle's coordinates having already been determined by Richer in 1671 when he departed for French Guiana); in 1681 they worked along the Channel; in 1682 they worked separately, Picard to Saint-Malo, Mont-Saint-Michel, and Caen, La Hire to Calais and Dunkerque. In 1681 two other persons determined the coordinates of Dieppe and Rouen. Picard and La Hire recorded their observations and some stories relating to their travels in the *Mémoires* of the Academy; three manuscript copies of a map of their Breton itinerary of 1679 have survived in the Service Historique de la Marine.[15]

To Picard, a coastal survey was only the first step toward completion of a map of all France. On 8 February 1681 he presented a memorandum on this subject to the Academy. Because those parts of France that had been surveyed were often separated by parts that had not been covered, the problem remained of assembling the surveys. This could be compared to fitting together the pieces of a puzzle whose overall dimensions were not known. Picard proposed that the length and breadth of France be determined first; then detailed field surveys could be fitted into a framework (*chassis général*) with little difficulty. To construct this framework, a network of triangles covering the entire country should be made, beginning with a chain from Dunkerque to Perpignan, approximately following the meridian of Paris. Completion of this chain would permit even more accurate measurements of the length of a degree. Other chains could then be made along France's coastal and land frontiers, which could then be connected to the meridian chain by chains following parallels of latitude (called in French chains perpendicular to the meridian).

Picard, however, died on 12 October 1682. On 12 June 1683 the Academy learned that Colbert had decided that Cassini should extend a chain of triangles the length of France along the meridian of Paris. Cassini, having shared Picard's work, enthusiasm, and vision, was fully capable of launching the enterprise; their efforts, after all, had been collaborative, not competitive. The method to follow having been determined, execution depended upon qualified people and adequate support. Three days later, Cassini read an outline of the project to Colbert, and on 29 June and 3 and 10 July he conferred with colleagues. To speed this work, which would involve correcting linear distance for the slope of terrain, Cassini proposed that two separate teams operate simultaneously. That summer Cassini led the team working toward the south and La Hire the team toward the north.

When Colbert died on 6 September 1683, the project of mapping France according to scientific principles and based upon scientific observations had therefore just begun. Although the method had been elaborated and tested, little fieldwork had been completed, and only since 1679 had the project appeared to gather momentum. Years of work lay ahead; how many, nobody knew. As a royal body, the Academy could pursue such work vigorously if that was the will of the king and of its patron-minister; but if they opposed such activities, which required so many men and so much money, there was little the academicians could do on their own to advance the project. Under Colbert cartography had been favored. François Le Tellier, Marquis de Louvois, who succeeded Colbert as patron-minister of the Academy, apparently evaluated the need for better maps differently, because Cassini was not authorized to conduct operations in the field in 1684. At the end of 1683 the chain of triangles reached from Mont Cassel in the north to Montluçon near the Massif Central.

When Louvois died in 1691, Cassini renewed his hopes. On 1 September he brought a project to the Academy calling for further measurements of the meridian and observations of latitude. The academicians did not show much interest, perhaps because in a time of war political conditions were adverse. Nevertheless, there was still something useful to do. The fieldwork undertaken between 1670 and 1683 had proved conclusively that France's Atlantic and Mediterranean coasts did not extend as far westward and southward as conventional maps showed. In February 1683 someone, possibly La Hire, had presented the Academy with a manuscript map that had the corrected outline superimposed on a base by Sanson. (This was, incidentally, the first map to show the meridian of Paris.) The Sanson map gave the area of France as 31,657 square leagues, the Academy map, as 25,386 square leagues. This map was engraved and printed in 1693 (fig.1).[16]

Fieldwork along the Paris meridian was resumed in 1700. Cassini I, joined by his son Jacques (Cassini II, 1677–1756), Giacomo Maraldi, and Couplet, reached Collioure, near Spain, the next year. Then the War of the Spanish Succession (1701–13) forced an interruption. Cassini I died in 1712.[17] In 1718, in the optimistic mood following the end of hostilities and the death of Louis XIV, Cassini II, Maraldi, and La Hire resumed fieldwork with nineteen new triangles between Montdidier and Dunkerque. The chain from north to south begun by Picard in 1679 was finally complete. In his report, Cassini II passed judgment on existing maps of France, comparing them against the findings of science for accuracy in determining the correct coordinates for the cities and natural landmarks of France. The first stage in making a modern, accurate map of France was over.[18]

Clearly, work on such a map had advanced more slowly than anyone had anticipated. That the French accomplished so little in fifty years does not detract from the fact that they were the only people to accomplish anything at all in geodesy at that time. At that rate, however, centuries would be necessary to conclude the superstructure of triangle chains Picard had planned to lay across France. If a map of France were to be made rapidly, many more workers would have to be trained and new techniques developed to verify and coordinate their work. Men, institutions, and money—far more than ideas—regulated the rate of progress. The national map survey had been launched in France under Colbert by people who perhaps under-

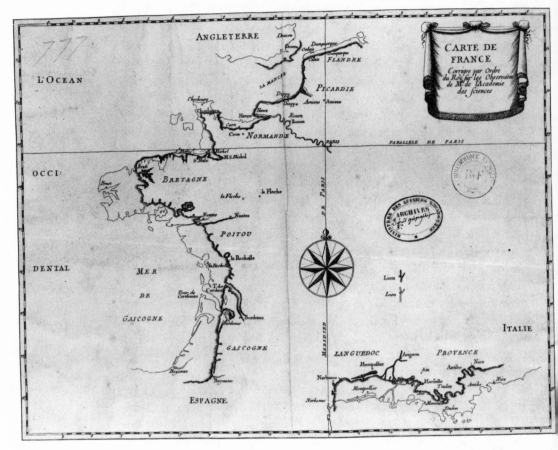

Figure 1   Traditional and corrected outlines of France compared. Engraved in 1693, this map showed the traditional cartographic outline of France in a thin, crisp line and the geodetically corrected outline in a darker line accented by coastal shading. "Carte de France corrigée par ordre du Roy sur les observations de Mrs de l'Académie des Sciences." 1:4,120,000; 36.0 cm × 26.6 cm. Phot. Bibl. Nat., Paris, Ge. DD. 2987-777.

estimated the effort, time, and money this venture required. Now that the difficulties attending upon it were fully known, under what conditions would French scientists and statesmen decide to continue the survey?

1730–50

When the state decided to resume cartographic fieldwork in the 1730s, Cassini I's son and grandson were conveniently available to supervise and complete the national map survey. Yet the Cassini dynasty in astronomy and cartography, like the national map survey itself, nearly did not survive. Jacques Cassini's marriage at the age of thirty-four to Suzanne-Françoise Charpentier de Charmois only confirmed the family's social ascension. Jacques combined two careers: an astronomer, he was also a *maître-ordinaire* of the Chambre des Comptes, a magistrate, and a *conseiller d'état*. His marriage on 7 April 1711 to a family of the *noblesse de robe* was celebrated by such dignitaries as the intendant of Paris and the *Prévôt des marchands*. Their first child, a son, was born sometime between 16 and 26 May 1712,[19] but this infant figures in no genealogical tree of the Cassini family. He died, presumably before the birth in 1713 of the Cassinis' next child, Dominique-Jean, who also became a *maître-ordinaire* at the Chambre des Comptes. Only the third son continued the family's work in science. César-François Cassini (also called Cassini III or Cassini de Thury), was born in 1714. Another son, Dominique-Joseph, born in 1715, pursued a military career; he was promoted to *maréchal de camp* in 1767. Cassini I died at the age of eighty-six, his son Jacques at seventy-nine. Cassini de Thury lived to be seventy, Dominique-Joseph died at the age of seventy-five, and Dominique-Jean at sixty-six. Given the risks attending infants and children, we are not surprised that Cassini I's first grandchild died in his first year. What is remarkable is that so many in the family survived to old age.

This is a fact of some importance for the history of cartography. Of Jacques's four children, only one became a scientist. The death of a single individual—Picard or Colbert—could jeopardize a large, long-term cartographic project. In the 1710s as in the 1680s, the institutions supporting cartography had not yet begun to recruit scores of practitioners, nor had they yet multiplied their connections throughout the French bureaucracy, drawing strength as a tree does by branching its roots. The physical survival of the Cassinis, and the choice of at least one Cassini in each of four generations to become a scientist, helped give the national map survey in France a century of continuity.

Ultimately, the government's desire to advance public works projects was responsible for the resumption of the national map survey. When Philibert Orry (1689–1747) became controller general in 1730, he realized that the work of the Ponts et Chaussées, the state civil engineering corps, suffered from inadequate accounting control, lack of coordination, decentralized decision making, and insufficient documentation. He decided to create an office in Paris to coordinate all public works projects, with maps as the instrument of control.[20] In 1730 Orry ordered Cassini II to resume fieldwork. Other units of the French state had mapping interests: the army's topographical engineers were already mapping sections of the country's frontiers; the army's fortifications engineers, organized in a separate unit in the years of peace after 1715, also acquired a mapmaking function; and in 1720 the navy created a hydrographic office, the first modern one to serve all mariners without regard to nationality. Each unit commissioned specialized kinds of maps. Sooner or later any one of them might have called for the resumption of geodetic fieldwork and the completion of the national map survey, but circumstances granted this responsibility to the Ponts et Chaussées.

By coincidence, at the same time as Orry acted, Cassini II and other scientists were returning to certain fundamental problems in geodesy related to the shape of the earth. The evidence Picard and Cassini had gathered on the length of a degree of longitude contradicted predictions based on Newtonian theory. Newton's work showed that the force of gravity varies from the poles to the equator because the earth is not a perfect sphere. When Jean Richer traveled to Cayenne in 1672 to study the effects of refraction and parallax in astronomical observations, he also observed that Christiaan Huygens's pendulum clock, which he had carried along, lost about two and a half minutes a day when close to the equator. Richer, Newton, and Huygens believed the clock was affected by the earth's shape, which they thought must be an oblate spheroid, flattened toward the poles. If the earth is an oblate spheroid, the length of a degree of longitude must increase from the equator to the poles; if it is prolate, that is, elongated, then a degree decreases nearer the poles. The fieldwork conducted by the Cassinis from 1679 to 1718 provided evidence that the earth is a prolate spheroid. Picard's measurement of a degree from Amiens to Malvoisine, north of Paris, was 57,060 toises; Cassini I's measurement of a degree in southern France was 57,097 toises; Cassini II's measurement from Paris to Dunkerque produced a degree of 56,960 toises. Cassini I was of the opinion that errors in observation rather than the shape of the earth accounted for these values; with better instruments, he believed, the truth could be established. Cassini II, however, became an advocate of the view that the earth is prolate.

After visiting London in 1728, Pierre Louis Moreau de Maupertuis wanted to explore the geodetic implications of the Newtonian system.[21] In 1729 and 1730, Maupertuis began work on the theory of spheroids, and he communicated papers to the Academy in Paris and the Royal Society in London in 1732 and 1733. Meanwhile, in 1730 Cassini I's memoirs on geodesy were republished, so that the differences between the Newtonian view of the earth as an oblate spheroid and the Cassini-Picard evidence of the earth as a prolate spheroid were drawn afresh. Therefore when Orry asked Cassini II to take up geodetic fieldwork once again, Cassini must have been eager to collect more evidence in support of his father's views. Orry was not a partisan of either faction in the shape of the earth controversy but only wanted accurate maps; his interest in geodesy was political and practical. Had geodesy lacked this appeal to such a man, the material and political support scientists needed to carry on ever more ambitious geodetic experiments would not have been so generously available.

Beginning on 1 June 1733, Cassini II and a team of assistants left Paris to begin work on a line perpendicular to the Paris meridian. (Measurement of the Paris meridian had been completed in 1718.) Pyramids four feet wide and eight feet high, or other large structures such as a kind of scaffold, were erected every ten kilometers if no suitable landmarks were present. Forests posed a special difficulty, since a thick stand of trees made sighting next to impossible. To get around some forests, the team decided to follow the Loire River, thus obtaining data useful to engineers at work on its dikes and embankments. In a regular correspondence, Cassini II kept Orry informed, but few of these letters have survived.[22] The team reached Saint-Malo, where its triangulation westward connected with Picard's earlier work. During the winter of 1733, a Cassini team prepared to continue the line of triangles perpendicular to the meridian eastward toward Strasbourg in 1734.

At the same time, Maupertuis began to plan a challenge to the work of Cassini in the form of two counterexpeditions. The Cassinis derived evidence for the view that the earth is elongated at the poles from their observation that a line of longitude between two degrees of latitude appears to diminish in length from the equator to the poles, but their data came only from the middle range of latitude. They simply inferred that the difference in length they measured was representative and continuous. A more conclusive test, Maupertuis argued, should involve measurements as near the equator and the poles as possible. The Academy lent its name and support to Maupertuis's proposal just as it had to the Cassinis' work, though in fact each group operated independent of the Academy. The Academy served to verify and authenticate the work submitted to it, but material and political support for such costly research could come only from the crown itself. The voyages of Pierre Bouguer, Charles-Marie de La Condamine, Joseph Jussieu, and Louis Godin to Peru and of Maupertuis, Alexis-Claude Clairaut, Charles Camus, Pierre-Charles Lemonnier, and Abbot Outhier to Sweden required money, passports, ships, and supplies. The Swedish expedition left in 1734 and returned in 1737, but the Peruvian one lasted much longer, from 1735 until 1744, and quickly outlived the government's willingness to support it.

The story of these expeditions has been well told already.[23] Maupertuis, with the help of Anders Celsius, using a nine-foot telescopic sector made by George Graham of London (with silver cross hairs fixed on springs, thereby held in constant tension and nullifying the effect of temperature changes), undertook to measure a degree at the northern end of the Gulf of Bothnia under extremely difficult conditions. In summer the scientists were afflicted with bloodsucking flies in thick, visibility-reducing swarms, and in winter they worked in extreme cold and with minimal light, yet they managed to measure a baseline of slightly over 14 kilometers (7,400 toises) and a network of triangles over 107.24 kilometers. The baseline was measured twice, and other geodetic measurements were checked with tests of the pendulum (repetitions of Richer's experiments in South America). Converted into metrics, Maupertuis's measurements gave the length of a degree as 111.094 kilometers, more than half a kilometer longer than the length of a degree in France. Thus he felt confident in announcing that the earth is flattened at the poles, as Newton predicted, rather than elongated. The Peruvian expedition extended a line of triangles for three degrees in a high, inaccessible area where the natives were often hostile and the weather was an even greater threat. As in Sweden, the French in Peru would have earned fame had they simply survived their adventure. Yet in such hostile environments they performed delicate experiments as carefully as if they were in the garden of the Paris Observatory. Bouguer made the discovery that in the mountains the swing of the pendulum was slower, even after altitude (which affects gravity) had been taken into account. Bouguer's observation that the density of rocks appears to vary was the first sign that the earth's crust is not uniform, and hence that the earth's gravity varies from place to place. (The difference between an observed value of gravity and the mean value is called the Bouguer anomaly.) And of course the geodetic measurements themselves had to be corrected to read at sea level, a laborious task of calculation to which Bouguer was equal. Bouguer and La Condamine made separate sets of triangles; Bouguer's provided a measurement of a degree at the equator of 56,746 toises, or 110.598 kilometers, La Condamine, of 56,749 toises, or 110.604 kilometers. Taken together, the Baltic and Peruvian expeditions provided conclusive proof that a degree increases in length from the equator to the poles, thus validating the image of the earth as an oblate spheroid (fig.2).

While the two French teams were abroad, the Cassinis continued their work in France, extending a line west of Paris to Brest in 1735 along a different route from the one taken in 1679 and 1733. The importance of Brest, Nantes, and Saint-Malo in French maritime affairs was sufficient to justify such extraordinary efforts to ascertain their precise location as well as to map the outline of the coast. In 1736 another line

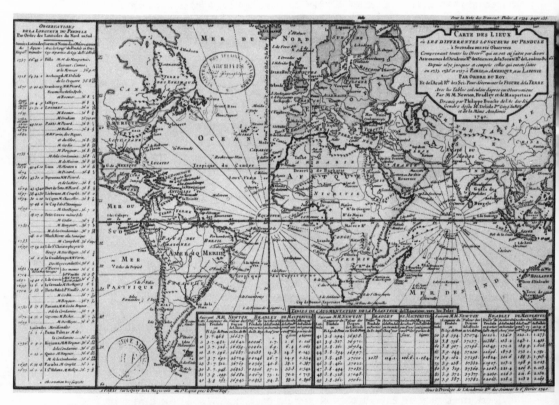

Figure 2 Sites whose geodetic coordinates were verified between 1670 and 1740. Engraved by Buache in 1740, this map is accompanied by a table on the left listing places where scientists measured latitude from the 1670s to the 1730s. The bottom table gives the values for the effects of gravity on the pendulum as a function of latitude. The map itself simply displays the world. Note how well the Saint Lawrence River and the Great Lakes are depicted compared with the southeastern part of North America. "Carte des lieux ou les différentes longueurs du pendule à secondes ont été observées. . . ." 1:80,000,000; 36.5 cm × 21.0 cm. Phot. Bibl. Nat., Paris, Ge. DD. 2987-136.

was extended west to the coast from a point on the Paris meridian sixty thousand toises north of Paris. Enough information was gathered to sustain the view that hydrographic charts of the French coasts were still considerably short of perfection. To acquire more data, a second north-south chain of triangles was extended in 1737 by Cassini III and Maraldi, passing through Cherbourg, Nantes, and Bayonne, while Cassini II and Nicolas-Louis de Lacaille covered the Channel coast between Saint-Valéry (Normandy) and Dunkerque. In 1738 a chain was extended from Bayonne to Antibes, permitting revision of Mediterranean charts; Maraldi went on to Nice, while Cassini III and Lacaille covered the Rhône delta in greater detail.

When Maupertuis announced the results of his expedition, the effect on the Cassinis was a remarkable demonstration that science is not a religion, divided into sects each trying to monopolize the truth and invoking the power of its own saints and relics. Cassini II did not disavow the viewpoint he had always sponsored, preferring to let his son, Cassini III, take responsibility for fieldwork. Cassini III neither conceded the debate nor attempted to defend an untenable position. Instead, with Lacaille, he set about making additional geodetic observations in France that would either support or disprove the work of the French abroad. They remeasured an arc of meridian south of Paris and then three baselines, in Bourges, Rodez, and Perpignan. To verify latitude and longitude, they made astronomical observations in the field, checked against observations in Paris. In 1740 they worked north of Paris. They began to suspect that Picard's measurement of his original base, from Villejuif to Juvisy, was itself the cause of subsequent errors, striking evidence of how an error in one part of a triangulation scheme can work its way through an entire network. Cassini III and Lacaille remeasured the baseline five times in 1740 and found that Picard's measurements exceeded the baseline by five toises.

Had Picard's measurements been more accurate and had they conformed with Newtonian theory, the French expeditions to Lapland and Peru and across France would have been unnecessary. Yet the expeditions alone provided convincing proof— the first experimental evidence, really—that nature behaved as Newtonian theory predicted. Furthermore, the expeditions demonstrated the feasibility and value of global scientific ventures and served as the inspiration and model for others later in the century. The oblate/prolate debate had been salutary for science in general and for cartography in particular. In a way, it was fortunate that Picard had made a mistake.

By 1740, four hundred principal triangles and eighteen bases had been measured in France; by 1744, when the Cassini team completed its work, there were eight hundred triangles and nineteen bases. Clearly, eight years of fieldwork and calculations had not exhausted the team, which faced four more years of hard work. The task undertaken between 1740 and 1744 was if anything more difficult and arduous than the work of the preceding eight years. The areas of easy access had already been mapped. What remained were the smaller villages, often in remote areas, difficult to reach, where amenities were few and the presence of the academicians was often unwelcome. The physical effort of transporting delicate equipment and supplies for daily life back and forth across terrain where footpaths were often the widest routes and often did not go where the academicians wanted—not to mention the long periods of waiting for the weather to clear—placed a premium on group loyalty and trust, sheer physical endurance, and a will to overcome unexpected and difficult circumstances. In the winter, when not out in the countryside, the scientists spent long hours performing calculations and making plans for the next summer's work.

There were also unexpected compensations. Cassini III noted that in the Auvergne cheese making was crude, dirty, and poorly understood, and the cheese itself was mediocre, tasting of suet. By contrast, in 1739 he came upon a Swiss cheese maker on the Mont d'Or (Jura), instructing peasants in the art of making Gruyère, whose product he found excellent. The Cassinis watched him work often and admired his attention to detail.

But friction with peasants was more often the norm. Peasants were suspicious of the academicians, first because they could not understand and may have been afraid of what the scientists were doing, staring at the stars or sighting across fields with fantastic instruments, and second because they often believed the scientists' work would result in higher taxes. It was hard for uneducated people in small rural communities, whose only visitors were peddlers and priests and tax collectors, to believe that the academicians were engaged in a purely innocent activity. The scientists carried letters of state explaining their mission and requiring local authorities to cooperate, but they had few defenses against stubborn and shrewd people to whom such documents meant little or nothing.

On one occasion even a well-educated government officer believed the scientists were merely posturing.[24] In this case he believed they were spies or agents provocateurs. During the summer of 1743, the Cassini team was at work in the Vosges Mountains at the very moment when the French feared an uprising in Alsace and Lorraine, given the unsettled international situation at the time. The local inhabitants had reported seeing eight richly dressed individuals and their servants working with telescopes in an area known to be populated by Anabaptists, a suspect group. The scientists erected large scaffolds ten meters tall on mountaintops near Rosheim, illuminated at night with large fires when it was necessary to make geodetic and astronomical observations simultaneously. Some army officers thought this was a way of sending signals for a revolt to begin. A subordinate official in Strasbourg reported all this to Paris, adding that the scientists worked on Sunday and left helpers behind or paid local people to watch over their structures and light fires at the proper moment. The year before, he reported, the strangers had been in the same area and had left behind a pile of wood and a stone marked with the initials J. G.; and one of the principals kept notes in a large notebook. In the mind of de Blinglin, the official, these were troubling reports. Anxious to appear diligent in his duties, he passed all this on to his superior lest he later be accused of having ignored some vital piece of intelligence. What a disappointment it must have been for him to learn through letters from Paris and Nancy that the authorities had sent these strangers to the Vosges on a mission. J. G. stood for J. Grante, one of the workmen. De

Blinglin made a feeble effort to redeem himself, suggesting that nevertheless the scientists were involved in a plot. The Cassini team probably never knew they were under military surveillance.

Meanwhile the preparation of the map of France for publication advanced. In an undated memorandum probably written in 1735 or 1736, Philippe Buache urged that the results of fieldwork be made available to engineers, as originally planned.[25] Buache (1700–1773) proposed the publication of tables of geodetic data or maps showing triangulation networks. Using these materials, engineers making more detailed maps of canals or roads planned or under construction could achieve a precise fit between the individual sections of a major public works project. Buache also designed a way to publish a map of France in many sheets. His sketch showed France divided into 116 consecutively numbered sheets (fig. 3).

In the end, the Cassini map of France was issued on 18 sheets, each covering considerably more ground than if an edition of 116 had been executed. A smaller

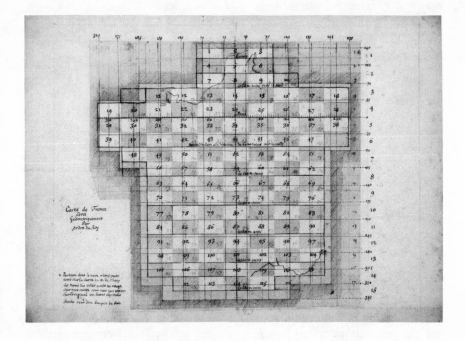

Figure 3   Outline map of France divided into 116 segments. The numbers written along the top and the right side are a scale; they correspond to distance from the meridian or the line perpendicular to the meridian. Checkerboard shading and bold numerals distinguish the 116 segments into which a map of France could be divided. This map, hand drawn by Buache about 1735, may have been made to suggest how the wealth of information the Cassinis were collecting could be assembled for publication. The Cassini survey of the 1730s and 1740s appeared in 18 sheets; the survey launched in 1750, in 180 sheets. "Carte de France levée géométriquement par ordre du Roy." Phot. Bibl. Nat., Paris, Département des Estampes, petit in-fol. Ve 6.

edition cost less and took less time to produce, and it was adequate for the needs of engineers. The Cassini team began to engrave and print as field surveys were completed, beginning in 1738. Several plates were revised during this period as additional information became available. As a result, the engraving and printing of the map were completed soon after field operations ended in 1744. The first charts drawn from field surveys were crowded with triangulation lines. By contrast, the engraved maps contained only the primary triangle networks. Engineers who needed to know the precise location of places and the distance between them could have used the maps even without triangulation lines. The triangle networks gave visual, graphic proof that the maps' content was reliable, in much the same way as architects sometimes expose structural elements to enhance the impression of a building's strength. In any case, the triangle grid became part of every map reader's mental image of France, as if the lines really existed in space (fig. 4). The general omission of information about landforms did not compromise the usefulness of the maps, not only because engineers needed such information in far greater detail than would ever be possible on these printed sheets, but also because the field survey teams had not surveyed landforms systematically.

The map finished in 1744 completely fulfilled the scientific and political objectives pursued by Colbert, Picard, and Cassini I between 1668 and 1683 and by Orry, Cassini II, and Cassini III after 1730. Yet this map has conventionally been represented, in error I believe, as a preliminary effort to the national map survey launched by Cassini in 1747–50. The second survey covered France in greater detail, in 180 sheets instead of 18. In 1744, however, Cassini III did not anticipate making another, more detailed map survey. We should therefore speak of two Cassini national map surveys of France, one finished in 1744, the other in 1788, each independently conceived and executed.

Had the Cassinis and the government thought of the work undertaken between 1733 and 1744 as preparatory to more extensive surveys, then the engraving of the first national map survey would have been unnecessary. There is persuasive evidence in the administrative reform of the Ponts et Chaussées, however, that the government considered the first national map survey an end in itself.[26] In 1738, when Cassini's survey was well advanced and engraving began, Orry established the corvée system of forced labor needed on large public works projects. The same decree (13 June 1738) that established the corvée also stipulated that engineers should begin drawing up uniform maps of highways to be built or repaired, taking account of the need to provide better communications in each region as well as between adjacent regions and between each region and Paris. In 1743, when geodetic fieldwork was nearly at an end and the national survey maps were almost complete, Daniel-Charles Trudaine (1703–69) was placed in charge of the administration of the Ponts et Chaussées in Paris. In 1744 he established a bureau de dessinateurs, a drafting room where the maps of public works projects would be made, coordinated, and collected, originally by a staff of three. In 1746 four more draftsmen were added, and in 1747 an additional six. The first head of this office was someone named Mariaval, but his performance

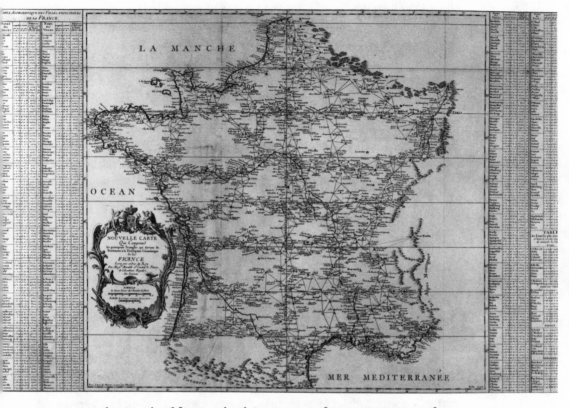

Figure 4 Triangle networks of first completed Cassini survey of France. It is apparent from this map that the survey teams concentrated on coasts, river valleys, and plains; they did not venture into the Pyrenees, the Juras, or the Alps, areas surveyed later by army engineers. The tables on the sides gave latitude, longitude, and distance from Paris for scores of places. Drawn by Giacomo Maraldi and César-François Cassini de Thury (Cassini III), engraved by Dheulland in 1744. "Nouvelle carte qui comprend les principaux triangles qui servent de fondement à la description géométrique de la France." 1:1,800,000; 91 cm × 59 cm. Phot. Bibl. Nat., Paris, Ge. BB. 565-A.

was unsatisfactory, and in 1747 he was replaced by Jean-Rodolphe Perronet. Perronet transformed the work of the drafting office into a training program for engineers, the nucleus of the Ecole Royale des Ponts et Chaussées. He instructed engineers to make a grid of all major locations in their districts using Cassini's maps and tables. Clearly, the eighteen-sheet survey printed between 1738 and 1744 met the needs of the civil engineers. It may be coincidence that the three steps outlined above—the corvée, the promotion of Trudaine, and the establishment of the drafting office— appear to match the progress of Cassini's geodetic and cartographic work, but I do not think so. Given that Cassini's work had been initiated by the Ponts et Chaussées and was intended to further its activities, these administrative steps were probably a deliberate effort to synchronize the reform of the civil engineering corps with the completion of the Cassini map. When the second Cassini national survey was almost finished in the 1780s, the civil engineers claimed it was useless to them, a claim that only reinforces my view of the relation between the first Cassini map and the administration of the Ponts et Chaussées.

Léon Gallois, in his 1909 article on the Cassini mapping activities before 1750, doubted that Cassini III had planned the second national mapping survey of 1747–88 by 1744, or even by 1747.[27] Gallois admitted that the first few pages of Cassini III's book of 1783, *La description géométrique de la France*, gave the impression that Cassini had worked out his second mapping survey before 1747, because the details of its execution were placed in the text before the account of the events leading up to the decision by Louis XV to launch the survey, as if those details had been decided before that decision was made. But Gallois also thought that forty years had confused Cassini III's memory; perhaps he believed later on that he had anticipated the second survey while the first had just been completed. But Gallois admitted that the appearance of continuity between the two surveys is reinforced because the book Cassini III published in 1783 seems to correspond to the text of a book he announced at the Academy's meeting of 13 November 1745 as being in press. Even Gallois believed that Cassini III merely postponed publishing it for nearly forty years. The truth is that Cassini III wrote a different book in 1744 from the one published in 1783, and he never published his earlier work. Surprised by the sudden turn of events in 1747, when Louis XV asked him to undertake a more detailed survey of France, Cassini III simply put his manuscript away. It has remained in the archives of the Paris Observatory.[28]

How can one be certain this manuscript is the one Cassini referred to in his lecture at the Academy in 1745? In his remarks, Cassini outlined the work he was preparing as follows: the first part was to be devoted to instruction in the methods used to overcome difficulties in geodetic fieldwork; the second would present solutions to various problems in practical geography; and the final part would be a summary. His manuscript "Le parfait ingénieur" gave an identical division in nearly identical language. Internal evidence in the manuscript confirms that it may have been written before 1745, but no later than 1746. On one page Cassini III added a footnote referring the reader with a more specialized background in geometry and astronomy to a book by Bouguer then in press ("que l'on imprime actuellement"). This could only refer to *La figure de la terre*, which appeared in the *Mémoires de l'Académie Royale des Sciences* for the year 1744, published in 1746, and appeared separately in 1749. Bouguer wrote it upon his return from Peru, in competition with La Condamine, to see who could publish his account first. Bouguer won. This manuscript of Cassini's, then, is unquestionably the text Cassini III's audience at the Academy looked for in vain in 1746 or 1747.

Cassini intended his manuscript to be read in conjunction with the set of eighteen maps already printed. He began his text with a clear exposition of the reasons for making a geodetically valid map of France. After reviewing the state of cartography before triangulation, when geographers estimated distance by the duration of travel, lacked sophisticated instruments, used each other's maps as primary sources, and lacked the funds to change their practices, Cassini introduced Orry as the decisive reformer. Cassini related that they agreed a geodetically accurate map of France could be made only by entrusting the entire job to a single team. After discussing at some length the nature of triangulation, Cassini presented the field surveys from 1735 in chronological order. From his preface it is clear that he intended the technical sections that followed for the average fieldworker, recommending that he pay attention to detail, check his work, and perform his observations with great precision. Cassini III thought of his text not just as a record of what he and his colleagues had accomplished (which he presented in his book of 1783), but as a manual of instruction that would allow individuals with simpler instruments and less education to make accurate surveys.

Cassini III believed that progress in cartography depended upon popularizing scientific methods. He concluded his preface with a statement he had made to the Academy in 1745 about steps that might improve French cartography. His recommendations confirm that at that time he had in mind no further cartographic work for himself. The geodetic matrix of France having been fixed, more detailed maps could now be made of small areas. The accuracy of such maps could be controlled geodetically by the maps and measurements already published by the Cassini team. Cassini cited as examples Abbot Outhier's maps of the dioceses of Sens and Bayeux, the maps of Provence by the Société Royale des Sciences de Montpellier, the maps of the royal forests, seigneurial maps of estates ("presque tous les seigneurs ont des cartes exactes de leurs terres"), and army maps of the frontiers.

Perhaps Cassini III was influenced by Jean Baptiste Bourguignon d'Anville's attempts to popularize mapmaking techniques in the 1730s and 1740s. D'Anville had implemented a suggestion François Chevalier made to the Academy in 1707. Disappointed at the time and money needed for geodetic mapping, Chevalier suggested that amateurs be recruited and instructed to make maps in the following way. First, a professional should make thirty or forty base maps in the form of circles and send them to as many people in as many areas. Each observer should place himself in the center of a circle, covering a total of two leagues and divided into eight parts. Using tables of the sun's declination and his own observations, the observer should establish points of sunrise and sunset relative to his position and trace the path of the sun across the area within his circle. Distances could then be marked with concentric circles at quarter-league intervals. Upon filling in his map with detail, using the concentric rings and the path of the sun as controls, he should return it to the professional. That person would then make a finished map out of the preliminary ones. Chevalier thought that in this way maps of dioceses could be made rapidly and with minimal effort. D'Anville tried to put this plan into effect. Invited in 1720 by Henri-Ignace de Brancas, bishop of Lisieux, to draw a map of that diocese, d'Anville distributed base maps and a text of instructions to the curates. Two published texts of instructions as well as two of the base maps completed by an observer in the field have survived (fig. 5).[29] But the questionable accuracy and limited verifiability of these maps were obvious impediments to the widespread use of such methods.

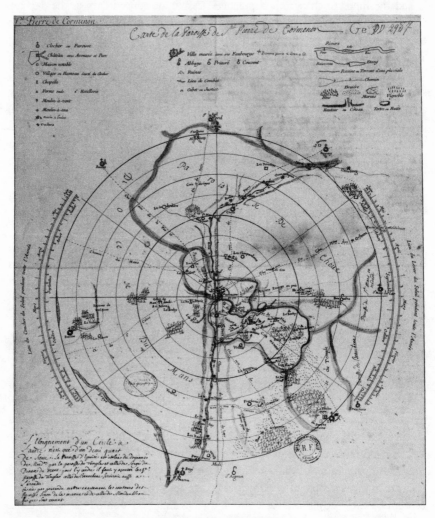

Figure 5 Parish of Saint Pierre de Cormenon. This map is oriented with north at the top. The circles represent distances of half a league; the mapper stood at the center. Drawn in the 1720s. 47 cm × 51 cm. Phot. Bibl. Nat., Paris, Ge. DD. 2987-1209.

Cassini may have thought his own book of instructions represented a significant and critical advance—when taken with his geodetic work—over Chevalier and d'Anville.

Cassini III believed that as knowledge of sound geodetic practices spread across France—certain to occur once his text was published—the quality and quantity of local and regional maps would generally increase. Further work by scientists was unnecessary, in his view: the private sector and local government together could sustain the preparation of more detailed maps of France. Cassini's manuscript gives us to understand that he intended the map finished in 1744 to provide the scientific authority for more detailed maps of France. These, however, were not to be made by Cassini III from fresh surveys, but instead would be made by other individuals and groups using the techniques and standards defined in his work. Once local and regional maps had been made according to these standards and techniques, they could be fitted together to compose another map of France at greater scale and in more detail. The conceptual and technical differences between what Cassini III had in mind in 1745 and the map he later went on to direct were enormous. The first national mapping survey of France was complete in 1745. Yet by 1750 Cassini III had begun a new mapping project, one that would bring him his greatest fame.

1750–93

In 1744 Cassini III articulated a vision of private and public agencies sponsoring scientifically valid mapping ventures in the absence of any central, coordinating unit. But his vision was naive. A second Cassini national survey mattered more to the welfare of cartography in France than Cassini III himself had thought. It acted as a catalyst. Several projects for collecting and analyzing data on natural resources and economic and social conditions relied upon the sheets of the second Cassini survey.

The second Cassini national map survey was a matter of national pride. As a publishing venture it has few peers in the history of cartography. Even today people admire the quality of engraving and enjoy looking at the features of France two centuries ago; it is one of the best-known artifacts of the eighteenth century. But in the eighteenth century its reputation as a scientific document took precedence over its graphic qualities. Few people then purchased sheets and still fewer bought complete editions, yet many were aware of its existence. People who never saw even a portion of the Cassini survey nonetheless knew that France had been mapped in unprecedented detail and accuracy. The Cassini map represented the conquest of space through measurement.

Its most lasting impact on cartography was perhaps the proof it gave by example, that such a vast enterprise could be undertaken and successfully concluded. Consider the difference between attitudes toward mapping in the 1670s and 1740s. In the first instance mapmaking projects were launched by people who underestimated the effort, time, and money they would require. In the second instance they were promoted by people who knew full well what difficulties they might confront but were not intimidated by the prospect. The very challenge was all the more worthwhile in their eyes for being so great.

The second map survey originated in the army's need for better maps. In 1744 the army's corps of geographical engineers was given a new status and added responsibilities. During a campaign in the Austrian Netherlands in 1744–45, French engineers attempted a systematic description of the right bank of the Scheldt from Audenarde to Termonde and of the Deudre as far as Alost, at a scale of 1:14,000.[30] As discrepancies between their maps became apparent, de Regemorte, head of the fortifications office in Paris, urged that baselines be measured. Since this task exceeded the abilities of the engineers in the field, de Regemorte proposed to the Comte d'Argenson that a triangulation network be composed for all the area the engineers were to cover. Cassini III, recruited for this purpose, left Paris on 1 April 1746. He measured two bases and two chains of triangles that were joined at a third base between Ghent and Bruges. At this point the French connected the triangles they had already measured from the Pyrenees north with the first chain laid out by Snel over a century before. When Cassini was about to return to France, he wrote to de Regemorte (16 October 1746), promising to bring him various maps he had made. Cassini suggested de Regemorte use this work to coordinate and adjust the work of engineers. Perhaps thinking of the map of France he had just finished, Cassini also wanted to publish a revised map of Flanders. But that was impossible, for as de Regemorte wrote to Françqis Masse, one of the army's most experienced engineers, d'Argenson would consider publication of even the smallest fragment a crime. Cassini's work for the civil engineers was supposed to be widely diffused, but his geodetic survey of Flanders was a military secret.

In a review of his victorious troops, Louis XV decided to compare several of Cassini's maps against the actual terrain. The king found Cassini's maps so accurate that he was inspired to order a map of France made to the same level of detail. Cassini III was instructed to take charge of the project and to approach Jean Baptiste Machault d'Arnouville, then controller general, about financing. Modestly yet truthfully, Cassini III admitted later that he had no idea that so momentous a request would result from his interview with the king.

Unlike the first national survey completed in 1744, the second survey would neither provide any immediate benefit to an agency of the French state nor offer scientists an opportunity to undertake experiments or test hypotheses. Cassini III, always the realist, knew that the success of this new enterprise would be determined almost entirely by organizational and logistic matters. Herein lay his chance to make another original and most significant contribution to cartography. The first survey had demonstrated the importance, validity, and practicality of a national geodetic survey, but its organization had depended too much on the support and talents of a few individuals to constitute an easily replicable model. The first survey did not imply an extensive, permanent mapping function for scientifically trained experts in government, but rather involved a brief role for the state followed by a longer period when various public and private mapping interests could pursue their separate, more specialized interests within a common framework. The second survey, however, held the potential for making mapping a permanent and continual activity of government. Cartography could then become a routine public activity, comparable to tax collecting or the maintenance of a standing army.

Cassini III began by translating various technical criteria into bureaucratic terms and categories.[31] He fixed the scale of the maps at 1:86,400; each sheet would represent an area 40,000 toises long by 25,000 toises wide (80 km × 50 km). Cassini estimated that 180 sheets would cover the country. Placed side by side, the sheets cover an area eleven by eleven meters. Because much of the basic geodetic work had already been accomplished, he devised a scheme whereby engineers, using less accurate instruments than would be appropriate in measuring primary triangles, could cover ground paying close attention to the detail of landforms omitted in the first survey. The engineers were to work at a large enough scale in a small enough area so that Cassini claimed any inaccuracies that appeared in their work would be detected by supervisors and would not accumulate to affect the accuracy of other maps. No scientific problems needed to be solved.

Each engineer was to keep two records. The first, containing the names of locations, observed angles, and any other relevant information, was to be signed by parish priests and the gentry. The second, containing geodetic information connecting localities to the primary geodetic triangles, was to support the construction of the map itself. There were thus to be two controls on the fieldwork of engineers, one by local dignitaries, the other by supervisory staff in Paris. In 1755 Cassini ordered that if a certificate from a squire or curate was missing, or if a supervisor failed to correct an important error, then one hundred livres would be withheld from the wages of the engineer responsible. Indeed, that sum was usually withheld as a bond of good performance when an engineer left for the field, to be paid only if the work was satisfactorily finished—a sign that Cassini did not always expect the best results. Later Cassini, complaining of inaccurate fieldwork, ordered the engineers to measure all three angles in each triangle and to measure all roads and four or five points on the boundaries of cities as well as the extremities of all villages longer than four hundred toises. The engineers were to keep three additional record books, one noting the angles and sides of all triangles, including the distances from observing stations to the meridian and to the nearest line perpendicular to it, another listing parishes in alphabetical order with distances from the primary geodetic matrix, and a third listing distinguishing features of each parish, again alphabetically. These lists and record books were designed to ensure accuracy and uniformity—no easy task considering that at least eighty-three individuals worked on the maps at one time or another.

The enterprise, Cassini estimated, would cost a lot of money. Financial details almost never appear in the letters, reports, and books written by Cassini II and his son about their mapping activities between 1733 and 1744, but they are abundant for the second map. The first Cassini map succeeded by virtue of the scientific abilities and hard work of the Cassini team, its close personal ties to high levels of administration, and its obvious scientific and practical significance. The second succeeded because it was well managed. The first map might have failed, but not for lack of money. Funding problems could have killed the second map survey at almost any time between 1747 and 1788.

Cassini began by assuming that ten maps could be made each year. At that rate the entire project would be completed in eighteen years. Each map would absorb the energies of two engineers. At the cost of four thousand livres per map (including engraving), annual expenses would amount to forty thousand livres. Machault, the controller general, thought this sum was not too great annually and even offered to authorize a larger subvention if the map could be completed in fewer years. But Cassini was unable to form survey teams quickly. Most of those who had collaborated on the first map either were not inclined to work on the second or had found more remunerative and secure positions, for example, in the army's corps of geographical engineers. New people had to be recruited, instructed in surveying techniques, mathematics, and draftsmanship, and given sufficient opportunity for practice to develop a consistent, uniform style. By 1756 only two maps—of Paris and Beauvais—were published. In that year Machault's successor, Jean Moreau de Séchelles, suspended payments, since funds were needed more urgently to prosecute a new war, which had just begun and was destined to last until 1763.

Having at last formed enough teams to increase the rate of activity, Cassini tried to persuade de Séchelles to change his mind. Failing in that, he did not despair but rather applied a solution that the Bourbon kings themselves had used often enough to initiate and underwrite a project: he formed a company and sold shares to raise capital. (It may well be that this solution was proposed to Cassini by de Séchelles or by one of Cassini's court friends, even though he claimed it was his own idea.) As Cassini noted, many of the leading members of the nobility showed an interest in his mapping activities. Madame de Pompadour, several ministers, the Prince de Soubise and the Ducs de Bouillon and Luxembourg were among the first subscribers. In a week Cassini had found fifty shareholders at 2,400 livres each, as many as he needed. Camus, Etienne Mignon de Montigny, Georges Louis de Buffon, La Condamine, and Marc Réné de Montalembert from the scientific community offered to help Cassini in running the company. Early on, Perronet took charge of training the technical staff and surveyors. When Camus died in 1768, Perronet, who along with Cassini was France's most experienced coordinator of cartographic surveys, was made a director of the company. One of the farmers general, Borda, was the company's treasurer.

According to documents drawn up by a notary on 10 August 1756, the king made a gift of all existing materials, instruments, maps, and working documents; the number of engineers was increased from Cassini's original figure of twenty to thirty-four; and annual expenses were fixed at 80,000 livres, of which 56,000 were allocated to salaries and 24,000 to printing costs and supplies. To assist the company in its early years, the controller general granted a subvention of 150,000 livres, which the various généralités were to reimburse, on the grounds that they would benefit from the maps when finished. The number of copies of each sheet was fixed at 2,500. The average cost had meanwhile risen from Cassini's original estimate of 4,000 livres a map to something closer to 4,700 livres. Since each copy was to be sold for 4 livres, the sale of all 2,500 copies of each map would yield 10,000 livres, and of all 2,500 of all 180 maps, 1,800,000 livres, a sum considerably greater than production costs. In fact many maps, for example, of sparsely populated areas, sold very badly by the sheet, since few people were interested in them, whereas demand for maps of other areas, such as Paris, exceeded production. Someone subscribing to a complete set could pay 562 livres in five installments or 500 livres in advance; purchased singly, 180 maps at 4 livres apiece would cost 720 livres. In other words, the company existed to raise enough capital to sustain an enterprise that might have expenses in excess of 800,000 livres over a twelve-year period. In 1784, when Cassini III died and when the map was nearly complete, the company's accounts showed that income since 1756 amounted to 642,567 livres, 7 sols and expenses to 623,601 livres, 11 sols.

By 1760, 50 maps of the 180 projected were finished, all in the north-central part of France. Between 1760 and 1770, 38 were published, covering all of eastern France, central France north of Lyon, and western France east of Rennes and north of Poitiers. Forty-five more were available by 1780, covering Anjou, Poitou, the Dauphiné, and Provence. The Limousin, the Pyrenees frontier, and the region of Nice appeared between 1780 and 1789. All fieldwork but not all engraving was completed by 1789 (fig. 6; see also figs. 16 and 17). So much for finishing the second map in twelve years.

The only important scientific work to come out of the second Cassini survey was the project to link the observatories of Greenwich and Paris with a chain of triangles.

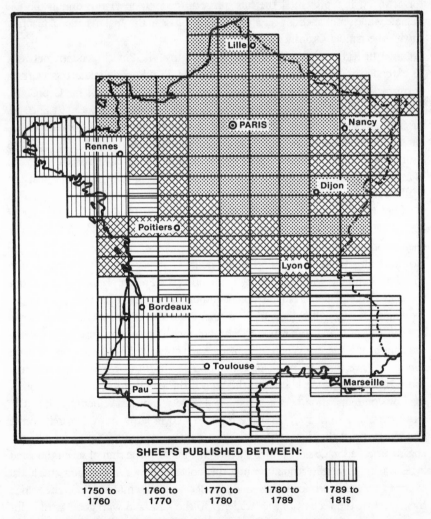

SHEETS PUBLISHED BETWEEN:

| 1750 to 1760 | 1760 to 1770 | 1770 to 1780 | 1780 to 1789 | 1789 to 1815 |

Figure 6 Cassini survey, dates of publication 1750–1815. Redrawn by Graphic Arts, Michigan State University, from Berthaut, *La carte de France*, 1:57.

Some unpublished letters in the Paris Observatory and some entries in the unpublished memoir of Jacques-Dominique Cassini (Cassini IV) supplement the printed accounts of this important step in the development of geodesy.[32] Alexander Dalrymple (hydrographer of the British East India Company) had exchanged information and collaborated on standardizing practices with various hydrographers in the French navy; French engineers had engaged in joint operations to map international boundaries; and voyages to observe the transit of Venus in 1761 and 1769 had been arranged jointly by France and England. The effort to link Greenwich and Paris geodetically, however, required more extensive preparation and intensive collaboration than any previous bilateral scientific or cartographic venture. In its own right it was unquestionably the most precise and refined undertaking of its kind, setting a new standard for accuracy. It represented the final phase in geodesy in France during the Old Regime while simultaneously serving as the first phase in the English national mapping survey, the famous Ordnance Survey.

Cassini III had proposed to various monarchs that the French geodetic network be extended to their respective capitals. He had already undertaken this work in Germany and Austria when the Royal Society of London, acting for George III, considered in 1783 and in 1784 approved Cassini's proposal in respect to England. Cassini's timing—in 1783 the American War of Independence ended—was excellent, for this project allowed the English to demonstrate their goodwill to the French in a subtle yet genuine manner. Joseph Banks wrote in rather good French to Cassini III on 11 May 1784, telling him that new instruments were to be made and that the English would begin by measuring a base near London.

Cassini III died on 4 September 1784. In 1785 Cassini IV showed an interest in continuing the project, but for the next two years nothing was done jointly by the English and French. This interruption was in fact necessary for Jesse Ramsden (1735–1800) to continue work on his great theodolite (finished in July 1787) and for the English to continue with preparatory fieldwork. (The French planned to use a *cercle répétiteur* by Etienne Lenoir, an instrument comparable to but different in design from the Ramsden theodolite. (Lenoir's instrument was the model for the one Pierre-François Méchain and Jean-Baptiste Delambre used later in the 1790s in measuring the length of a meter.) When in 1787 joint activities were resumed, the initiative appears to have come from the English, in response to their own mapping program, and was probably made known to the French through diplomatic channels rather than by communication between the Royal Society and the Academy or Paris Observatory. On 12 April 1787, Cassini IV received a cover letter from Armand, Comte de Montmorin at the court, transmitting two items that had been presented by the English ambassador. Cassini IV then submitted a budget. The sum of seven thousand livres was to cover the acquisition of a theodolite one foot in diameter and initial expenses for fieldwork for himself and for Méchain; Adrien Marie Legendre joined them later, at the request of the Academy to the king, and with good fortune for science. Charles Gillispie has written of Legendre's work: "In all the long history of surveying, the accounts that he published of those operations gave the only sophisticated analysis of the errors involved in computing the relations of lines and angles observed on a spherical surface by formulas from plane trigonometry."[33]

Fieldwork on both sides of the Channel began that summer and was completed late in the fall. This could have happened only if both parties were fully prepared and enjoyed excellent support services from their respective governments in the way of food and fodder, cooperation with local authorities, and dispatch boats to convey messages. The teams accepted Cassini IV's suggestion that they begin at their respective coasts, first connecting the two sides of the Channel with numerous observations and triangles, since this was the only phase of their project that had to be carried out jointly, and then connecting the coasts with the capitals later, each team working independently in its own country. Late in the summer of 1787, each team interrupted the triangulation network upon which it was at work and proceeded directly to the coast. The French went to Dover, where the two teams agreed upon their plan. The French and English erected signals in conspicuous places along both sides of the Channel, with the intention of illuminating each station according to a predetermined sequence for those on the other coast to observe. They had to deal with bad weather, contrary winds, fog and the technical imperfections of lamps; yet they were equal to the task, and nearly every observing session went off as planned. Anyone familiar with that area of the world can appreciate what they endured from the elements. By 13 November the French were ready to rejoin their English colleagues and crossed the Channel again.

Cassini IV, in his published version of their joint operations, acknowledged the hospitality extended to the French during their visit of three weeks in London but indicated that an extended description of their life in that city would be inappropriate in a scientific book. Such reticence is disarming and convincing, but it is also deceiving. In his unpublished memoir, Cassini confessed that his trip to London was only ostensibly for sightseeing. He had agreed in advance with Comte de Breteuil to go to London to try to get Ramsden to join the staff of the Paris Observatory or, if that failed, to get Ramsden or other English craftsmen to take on some French apprentices. In addition, Cassini traveled with an optician, one Sieur Carrochez, whose goal it was to examine William Herschel's telescopes and to learn some of the secrets of the English in optical glassmaking. The English did not entirely trust the French. The warm reception from the Royal Society that Cassini acknowledged in print he somewhat regretted in private. Two members of the Royal Society accompanied the French everywhere. Cassini stated that they might have seen more without their hosts, whose presence was a restraining influence on what the French scientists could learn. Cassini admitted that English instruments made him jealous, but his envy diminished upon reflecting that in England there were many observers and not enough instruments. Ramsden was willing to make an instrument for the French, but he was hesitant to take any French apprentices and certainly would not move to Paris. An English glassworker, formerly French and a Protestant, was willing to return to France if he could get his property back. The Baron de Breteuil said he could arrange it, but he left office in 1788 before he could do so.

The French had good reason to take such an interest in the work of Jesse Ramsden. In 1768, a year before his death, the Duc de Chaulnes had developed a method for mounting microscopes to a circle so that diametrically opposite points on the circle could be read, thus permitting a craftsman to mark the division of a circle into degrees more accurately than had been possible before. But he left no successor in France. In the meantime Ramsden dominated the field of precision instruments for geodesy. Ramsden's machine for dividing the circle was the first practical one of its kind, making "possible very rapid and highly accurate division of circular and rectilinear scales."[34] (Jean Charles de Borda had sent his instruments to England to be divided until 1783, when Lenoir began that kind of work in France.) He gained greater fame for his giant theodolites; the first was completed in 1787 and the second in 1790. They remained in use until 1862 and had a probable error of about five seconds over a distance of seventy miles. Cassini IV had taken steps in 1785 to create a model workshop at the Paris Observatory, and in 1787 he acted to establish a body of licensed instrument engineers, which actually functioned from 1788 to 1790. In France, makers of scientific instruments were not members of the Academy; in England, the most famous were members of the Royal Society. The social stature of the profession was important because the work of making instruments required much capital as well as state of the art skills in metallurgy and optics. Cassini understood the need to promote the profession in France, but the indifference and neglect of a generation was not dissipated until the Revolution. Gillispie did not miss the irony that "here at the end of the old regime, Cassini and Bailly, the most sycophantic of the scientific community in their attitude to aristocracy, should have been brought by circumstances to exemplify all unwittingly the democratic presence of Diderot's dictum animating the *Encyclopédie:* that in the interests of rationalized industry, artisans must be raised in the world and taught to have a better opinion of their own worth."[35]

Instrument making had become critical to accuracy. In going over their respective observations, the French and English discovered discrepancies. Cassini found that the largest French error in any one triangle for all three angles combined never exceeded four and a half seconds. The largest English error was a little over two and a half seconds. Nonetheless, the French too could be proud. They measured one trans-Channel angle twelve times in one day, twenty in another, and ten in a third, with a total of only six-tenths of a second in errors. In specific cases there were some differences between the English and French observations as great as twelve seconds, which occasioned an exchange of letters between Cassini and William Roy and some comments by Cassini in his book on their joint project. Naturally Cassini found fault with the methods of the English, but he also recorded some extenuating circumstances to explain their miscalculations. Each praised the other's instruments in private and in public. Using different techniques, they had come to within two-tenths of a second in their calculations of the difference in time separating the observatories of Greenwich and Paris, which is to say their respective longitudes. It is revealing of their ideas about science that Roy and Cassini agreed to publish their results and their differences, so that the public, seeing them, would trust that the work they had finished approached perfection, perfection itself being unattainable. Clearly, Cassini IV must have faced the future at the end of the 1780s with considerable satisfaction in his own achievements.

This prototype of the modern national survey was still conditioned by the institutional characteristics of the Old Regime. In no matter was this more obvious than the relation of the second survey to the governments of some of the provinces. Instead of reimbursing the state for their share of the 150,000 livres subvention, Burgundy, Bresse, Artois, Languedoc, Guyenne, Provence, and Brittany all opted to make their own maps, which eventually would be included in the national map survey. Some were very good; one excellent map, of Guyenne, has been carefully studied.[36] Brittany, however, made a political issue out of its project, thus delaying the work of the Cassini survey for that region. These provincial maps can also be viewed as the cartographic expression of the conservative, aristocratic reaction that marked the final decades of the Old Regime, impeding the reforms of the centralizing nation-state.

Linked to the Old Regime, the Cassini map suffered when it fell. The treatment of the map by the revolutionary government and its successors broke the bonds of trust that had tied Cassini IV and his forebears to France's rulers and ministers over 120 years. The departure of Cassini IV from cartography symbolized the end of an era. No longer would French cartography be so closely identified with a single name or so closely aligned with the fortunes of one family.

Cassini IV possessed the self-confidence and sense of superiority that an heir to a dynasty absorbs from the way others treat him. Others more gifted than he were more passionately committed to scientific work; others who had worked harder to obtain important positions and receive recognition were more reluctant to abandon their careers as ministers and even regimes came and went; others who had less wealth were more dependent upon their scientific work and administrative appointments for income. Cassini IV was in the uncommon position of being able to choose whether to remain a professional scientist.

As the course of the Revolution became more radical, Cassini's position and the great map so often called by his family's name were affected. It may be that Cassini's attachment to the monarchy and scorn for demagoguery may have contributed to his difficulties, but in all likelihood the elitist qualities of the Paris Observatory and of the map would have attracted the attention of politicians regardless of Cassini's personal views. Only if Cassini had been willing to accept changes desired by the revolutionary regime could conflict between them have been avoided. Cassini had professional as well as personal reasons for not cooperating. From March to October 1793 he opposed reforms at the Observatory on the grounds that unqualified people would be placed in positions of authority and that workers of unequal merit and responsibility would be given equal salaries. But the struggle to preserve an enclave of meritocracy in an egalitarian era was doomed. If he could not keep the Observatory as a fit place for research he preferred not to be associated with it, and he resigned on 6 September 1793. In the late 1790s attempts were made to reenlist Cassini's services, either at the Observatory or as head of the Bureau des Longitudes. However pleased Cassini may have been to be courted, he would not accept. Perhaps he could have returned to a position of eminence in state-supported scientific work had he only lost control of the Observatory. What made the breach irreconcilable was the state's treatment of the great Cassini map.

Until September 1793 the company retained its map. At that time, when Cassini resigned his position at the Observatory, the army's map division underwent a major reorganization as part of the government's attempt to centralize and unify all private and public map collections. The Cassini map, decreed the Convention, was to be transferred to the army, thus ensuring its survival and its utility for the state and preventing domestic or foreign foes from seizing it. The staff of engravers was retained and paid by the state. At that point 165 sheets had been engraved, 11 more were being prepared for publication, 1 was being drawn, and 4 remained to be edited from field notes.

The government did not intend to indemnify the company for its loss but only meant to pay for the value of the material and equipment transferred. The government defended its expropriation of the map on the grounds that it had really been started by the king and intended to serve the needs of the state; individuals had no right to profit from such an enterprise. The manager of the map, Louis Capitaine, was the only official of the company left to deal with the authorities, since Cassini had lost any room to maneuver politically and the former president and treasurer of the company had lost their heads. Upon receiving seventeen petitions, the Committee of Public Safety agreed to indemnify the company for the value of its loss, including capital gains, at the rate of 9,060 livres per share. But the sum was paid to only four of the fifty shareholders. A law of 24 frimaire year VI reduced each share to a value of 166 livres, which Napoleon, in a decree of 7 ventôse year IX reversed, acknowledging the state's debt and ordering that shareholders be paid. Meanwhile Cassini had submitted petitions and memoranda calling attention to the facts and claiming that a great wrong had been committed. He also made his case public in his published memoirs. In 1818, at the time the government was canvassing several departments about making another map of France, Cassini called for an investigation and requested that the map in its entirety be returned to the company. The government agreed to an inquiry; Pierre de Belleyme, head of the geographical section of the archives, acted on behalf of Cassini, and Pierre Jacotin, the colonel in charge of the topographical section of the army, acted for the government. The Council of Ministers decided that each copperplate should be assessed individually for wear and given a value in current francs, which turned out to be a sum of about 150,000 francs. Furthermore, Jacotin believed the state should retain possession of and continue to print the maps during the thirty or thirty-five years that remained before the next national map survey would be finished. Cassini responded that the value of the plates had been diminished since 1793 by the corrections and printings undertaken by the government, a loss for which he claimed compensation. But the government, rejecting Cassini's petition, divided the assessment into units of 3,000 francs per share. From then on the French state has retained possession and ownership. Cassini IV spent the last forty years of his life (he died in 1845) as a justice of the peace, a member of the general council of his department, and an enlightened administrator of farms and forests. One of his sons, Alexandre-Henri-Gabriel (1781–1832), started out in astronomy, studied law and served as a judge, and nurtured an interest in botany. He was the last Cassini to be admitted to the Academy.

The fate of the Cassini map after 1793 was conditioned by changes in the role of cartography in the bureaucratic workings of the state that were already under way before 1789. Cassini was of course frustrated by the loss of control over a project he and his father had supervised for forty years, and he had also lost money. He probably failed to realize, therefore, that the expropriation of the national map survey could be explained by the growing interest of the state in mapmaking and map collecting, which had its source in the professionalization of cartography, and the rivalry between groups of cartographers from various bureaucratic units at least as much as in the rhetoric and ideology of revolutionary politics. The evolution of the state's role in cartography before and during the Revolution is the subject of the next chapter.

Two

Cartography and the State in the Revolutionary Era

Cartography and International Borders before 1789

The topic of border surveys provides an overview of cartography in government in the half-century preceding the Revolution. In early modern Europe, borders were often easier to describe verbally than to represent cartographically. Territories were assigned to one state or another because they belonged to a given diocese or customs unit, for example, or according to some other nontopographic criterion that a surveyor could not depict; lawyers and statesmen frequently did not know which state a given village or district belonged to at the time a treaty was made.[1] Jurisdictions frequently overlapped, and nations often possessed enclaves, bits of land surrounded by territory belonging to another power. The appearance of a border as a continuous line on a small-scale seventeenth-century map simplified a complex situation. Far from being at all regular or consistent, in many areas the boundary had no clearly defined shape on the ground. Since the frontier was often negotiated with an eye toward the next war, and hence toward military advantages that might be gained from possession of certain topographical features, assumptions internalized in the treaty-making process included the likelihood that the boundary was not permanent. What had been acquired might be lost; what had been lost might be recovered. The determination of boundaries therefore depended on the good faith of both parties. For the most part, however, the people who lived in border regions knew which state they belonged to, even when they pretended to be confused so as to evade taxes. There was little incentive for governments to invest in better maps of their borders as long as these were so difficult to determine and so likely to change.

At Nijmegen, possibly for the first time, the desire to eliminate contentious issues about the sovereignty over border towns and about unclearly demarcated boundaries brought forward a proposal that rivers be used as frontiers, with each party giving up claims to land on the opposite bank. This principle was adopted more extensively at Ryswick, where the highest peaks were used to separate France from Savoy. And at Utrecht rivers were used as boundaries even if they divided towns. To avoid disputes over a boundary and to promote peace and stability in international relations, a map of the Flanders boundary was affixed to a treaty drawn up in 1718 between the emperor and the States-General. But this example did not set a trend.[2] Diplomats did not yet have maps of entire border regions to work with, but only maps of disputed areas. Most of the information they used was in the form of written reports and memoranda. Increasingly in the eighteenth century, treaties called for professional mapmakers from both parties to draw up a map jointly, to be signed by negotiators and annexed to the treaty as an official document. By postponing the mapmaking aspect of treaty making until after the written text had been concluded, diplomats gained time to study the border.

The Chevalier de Bonneval formulated a plan in 1747 to provide the French government with information about boundaries on a systematic basis before, during, and after negotiations.[3] Citing the confusing multiplicity of legal issues that complicated the determination of boundaries during negotiations, de Bonneval suggested that the government appoint an official to survey France's border regions and advise diplomats during negotiations. This person could prepare a large-scale map of a border region for use during negotiations so that the border could be marked on the surface of the map as diplomatic agreements were concluded. It would remain, then, for commissioners to travel through the border region, placing boundary markers along the line that had already been fixed during negotiations. De Bonneval wanted the map to be made before the treaty was concluded. But eighteenth-century practice continued along more familiar lines: the boundary was first described in the treaty text and then was represented cartographically on a map made by an inspection team working in the border region. The precise sequence was not critical either to the process of rationalizing border regions or to the use of maps in diplomacy. In the eighteenth century these were mutually reinforcing trends. The preparation of maps as part of treaty making had been exceptional before 1715; it became routine by 1789.

Whether in response to de Bonneval's suggestion that boundaries be routinely mapped in peacetime so that their geopolitical implications could be analyzed at length by military stategists and diplomats or to someone else's idea, the Ministry of War began to implement such a program of mapmaking from the 1740s. The army executed field surveys of France's northern and eastern land borders and western coastline until the Revolution. The diplomatic corps, however, operated without a geographical office until the 1770s. In the meantime, fortifications engineers from the army acted as the government's agents in making maps for treaties. Then things began to change. The Ministry of Foreign Affairs made efforts to acquire the largest map collection in Europe and to establish its own mapmaking staff.

Jean Baptiste Bourguignon d'Anville (1697–1782) was a map editor who designed new maps by collecting and comparing as many sources as possible. Scrupulous in his attention to detail, he corrected and reissued maps as new information became available, and he relied on correspondents for maps and information of all kinds.

Each important map was published with a critical essay on geography and cartography. His personal collection approached nine thousand items. The various units in the state bureaucracy that had assumed mapping functions maintained their own collections and did not often communicate or exchange material among themselves. The king had a collection of his own, but it functioned more as a great private collection than as a public reference source.[4] D'Anville's collection was comprehensive; the state's collections were not. A librarian of the Royal Library, Jean-Denis Barbié du Bocage (1760–1826), was a pupil of d'Anville's. Thanks to him, the d'Anville collection was acquired by the Ministry of Foreign Affairs.

The acquisition of the d'Anville collection is itself an interesting story.[5] In November 1772 d'Anville proposed to donate his collection of maps to the state. Apparently the Duc d'Aiguillon, minister for foreign affairs, agreed that in return, from 1772 on, d'Anville or his widow would receive a pension of 2,500 livres, and his daughter and only child, a pension of 800 livres. D'Anville then decided to retract his offer on the grounds that he did not yet want to lose his collection. So in 1773 Claude-Gérard Simonin, who was in charge of the ministry's archive, suggested that he try to compose a collection out of maps that could be found in various offices of the ministry. The office chiefs, reluctant to part with maps in their possession, did not admit to having any. D'Aiguillon then authorized Simonin to spend between 4,000 and 5,000 livres to build up a collection. Simonin gave the mission to Gaillard de Saudray, an official who occasionally made maps of foreign lands when on duty abroad. In one year, between assignments in Berlin and London, de Saudray spent 4,591 livres on 2,618 items.

At some point negotiations with d'Anville were resumed, for on 11 January 1780, in his eighty-fourth year, he agreed to give his collection to the king in exchange for a pension of 3,000 livres for himself and 1,000 livres for his daughter, plus the right to retain his apartment in the Louvre and keep his collection with him there until his death. D'Anville's son-in-law, Sieur d'Hauteclair, protested that d'Anville had underestimated the value of his collection, and hence of his daughter's estate. He asked for a lump sum of 30,000 livres or for the continuation of d'Anville's pension of 3,000 livres during his daughter's lifetime. The ministry recommended that d'Anville's daughter's pension be raised to 1,500 livres and that she receive 20,000 livres on her parents' death. In addition, d'Anville was to receive a tobacco case with the king's portrait on it, valued at 2,400 livres. The ministry also asked for 2,600 livres, the cost of making an inventory of d'Anville's collection (which had occupied several people for two months) and requested 2,000 livres for cases and bindings. The geographical office also asked for funds to buy a set of Cassini's maps from d'Anville, since the department lacked one. The final agreement, containing these provisions, was approved by d'Anville on 11 March 1780.

The inventory, numbering 8,788 items, was largely the work of Barbié du Bocage, who had begun work on it in 1779 at Charles Gravier de Vergennes's suggestion under the supervision of two of the ministry's geographers. Barbié du Bocage himself entered the service of the ministry on 9 March 1780. Then began the tedious work of checking all the titles and attributions of sources of the maps, as well as verifying scales. Barbié du Bocage thought that he could check some eight hundred to one thousand maps a month; in fact, the work was not over in a year, as would have been possible at that rate, but lasted until early 1782.[6] When d'Anville died on 29 January 1782, Barbié du Bocage took charge of moving the collection to Versailles (on 18 February, to be precise). In 1783 he began integrating the maps in d'Anville's collection with the ministry's other collection, whose existence he discovered only in that year. He had to make a catalog and restore maps in need of repair. In addition, in 1785 the intendant of Paris presented the ministry with some three hundred maps from Alexander Dalrymple.

In 1924 the Ministry of Foreign Affairs transferred most of the d'Anville collection to the Bibliothèque Nationale on the grounds that these items were no longer needed by diplomats in their work.[7] In World War II the Germans shipped almost all maps remaining in the ministry to Berlin—there were several thousand—to form the nucleus of a collection there, since the German foreign office lacked one, incredible as that may seem. The French maps were to be used in redrawing the boundaries of Europe. Only recently, one small part of that collection turned up in Poland and was returned to the Quai d'Orsay. The fate of the rest of the collection is still unknown.

The Ministry of Foreign Affairs acquired the d'Anville collection and simultaneously recruited a staff of cartographers. When Vergennes became minister of foreign affairs in 1774, he agreed to a suggestion put forward by Gérard de Rayneval (codirector for political affairs) to establish a geographical office.[8] On 1 January 1775 the Bureau Topographique pour la Démarcation des Limites was established. Its story has never been told; what follows is based upon archival materials within the ministry itself, principally its personnel files. Initially,the office was composed by J. A. Rizzi-Zannoni (chief), Chrestien de la Croix (deputy chief), Caffiéri, Brossier, and Vitry (staff), and two draftsmen.

Had the success of the venture depended upon Rizzi-Zannoni, it would have been a failure. J. A. Rizzi-Zannoni (1736–1814) was an Italian cartographer who had failed to secure a position in the hydrographic office through influence. Why the diplomatic corps accepted him is not known, but he enjoyed the protection of important people, and in 1771 he had been engaged to trace the boundary between Liège and France.[9] He had also been commissioned to make a map of the eastern part of the Turkish empire, so that affairs between Russia and Turkey could be followed more closely. Once in office, he became engaged in a lengthy and costly venture, the preparation of an atlas of France's boundaries. This would gather together all the information on boundaries contained in maps attached to treaties, maps that otherwise would have to be consulted separately. Whether Rizzi-Zannoni ever intended to complete the atlas or planned instead to drag it out for many years and incur heavy expenses that the government would have to cover cannot be determined. A prospectus for this work, drawn up by somebody else on 13 May 1775, suggested that the project would last three years and would involve annual salaries of 6,000 livres for Rizzi-Zannoni, 1,400 livres for Caffiéri and Chrestien de la Croix, and 600 livres for Brossier, as well as 1,000 livres for members of the Academy for astronomical observations and technical assistance, and 1,000 livres for a mathematician.

Even before Rizzi-Zannoni began work on the atlas, he ran into trouble with the government over the map of the Russian-Turkish conflict. Production expenses on that map already exceeded 7,000 livres.[10] To assist Rizzi-Zannoni, the king had lent him 5,000 livres for one year. Rizzi-Zannoni was to pay off that loan by selling 1,200 copies of the finished map, but his debts caught up with him first. Apparently he had not paid the engineers out of the loan but had pocketed the money (and spent it on high living in Paris). The engravers also complained about not getting paid. A well-known engraver, Guillaume Delahaye, evaluated the engravers' claims and found them modest, correct, and reasonable. The state then honored those claims (for 2,140 livres), attaching most of Rizzi-Zannoni's salary.[11] In negotiations in early 1776, Rizzi-Zannoni first suggested that the state acquire all copies of the map and recover the 5,000 livres by selling them, then suggested that he give up 500 livres a year from his salary for ten years. Neither proposal met with favor. The government proposed instead that 1,000 livres be garnished from his wages in 1776, 1,500 livres in 1777, and 1,780 livres in 1778 and in addition, that he give the ministry sixty copies of the map at an assessed value of 12 livres each. These terms were too much for Rizzi-Zannoni. On 14 July 1776, on the pretext of family business but really to escape his creditors, he left for Italy and remained out of contact with his friends in Paris for some time thereafter. To throw the government off his trail, Rizzi-Zannoni had taken the precaution of sending his trunks to Venice via Munich. In the end, the government seized the plates and printed 312 copies of the map in a futile attempt to settle this debt.

Clearly, Rizzi-Zannoni was not the man for the job. He was one of a few cartographers who used their skills to defraud their patrons. That he fooled some of the people some of the time only shows how well he played the game. Yet he still retained a reputation as someone important. In 1799, when the French occupied Italy, Rizzi-Zannoni offered to repay his debt by serving in the army. He also proposed that he be given a salary of 12,000 francs, that an additional 12,000 francs be given him to defray the costs of transporting himself and his household to France, that he be housed in Paris at government expense, and that he be given a percentage of the sales of his map of Italy. In return he promised to give the French all his instruments and materials. Remarkably, the French gave him 10,000 francs in cash and 20,000 francs for expenses, with the explanation that they wanted to prevent Rizzi-Zannoni from going to Prussia. In the end he stayed in Naples and made maps of uneven quality for the French. He might have served the French better had they allowed him to go to Prussia.

Following Rizzi-Zannoni's hasty departure in 1776, Jean-Sebastien Grandjean, a captain of dragoons, became head of the ministry's cartographic unit.[12] Grandjean had accumulated experience working on the collection of relief maps in the Louvre (1756–58), at Versailles keeping track of army positions on maps during the Seven

Years War (1759–63), in Africa mapping the coast near Gambia (1763–65), under Pierre-Joseph de Bourcet along the Savoy and Piedmont frontiers undertaking reconnaissance missions (1769–71), and surveying France's eastern and northern frontiers (1771–75). According to the ministry's records, his subordinate Caffiéri entered service in 1774; Vitry started in the hydrographic office before coming to the diplomatic service in 1775. These three, plus Chrestien de la Croix and Brossier, remained together into the revolutionary era. In 1793, when Grandjean retired, Chrestien de la Croix took his place; he in turn remained on the job until 1830. By virtue of the duration of their service, Grandjean and Chrestien de la Croix provided institutional continuity.

Several documents illuminate the procedures that the staff under Grandjean followed, first along the borders of Lorraine between 1778 and 1782, and then along the border with Spain from 1786 to 1793. Their work was not easy because no primary geodetic grid existed for most of the areas they covered. They had to undertake triangulation, often in isolated, rugged terrain. At first the ministry was unsympathetic to the problems of fieldwork, and in 1780 Grandjean, pressured by his superiors, had to write to his subordinates threatening them with the loss of their positions if they did not make rapid progress. Gradually the ministry became reconciled to the time and cost absorbed by cartography.[13]

Grandjean's memoranda for the Spanish border maps specified a scale of 1:14,400. From 1786 to 1789, two triangulation networks, one at a larger scale than the other, were laid down. Coverage of landforms was the responsibility of three engineers out of a team of four; the fourth handled geodesy. Given the enormous task of mapping a boundary in the Pyrenees, there were two four-man teams each from France and Spain; one French team was from the army, the other from the Ministry of Foreign Affairs. Originally, they were supposed to survey to a depth of two leagues on each side of the border. By 1789, only forty-four square leagues had been covered—about 10 percent of the total border area to be mapped. At that rate the map would require fifteen years to complete.[14] The French therefore proposed reducing the distance to be mapped on each side of the border to one league. At the same time, the French divided their tasks between their two teams, so that the army group started at the Atlantic and worked eastward, and the Foreign Affairs team started at the Mediterranean and worked westward.[15] This was, incidentally, one way of handling friction between the two teams, each of which claimed to be in control of the other. In principle, one Spaniard was to work with the French and one Frenchman with the Spanish. They were all still at work when French relations with Spain deteriorated in the 1790s. The Committee of Public Safety suspended Grandjean after he had been denounced for drawing the border with Spain in a manner prejudicial to France and then reinstated him in year III. Testimony given on his behalf confirmed that Grandjean was not responsible for where the border lay, but only for mapping it. Soon after he resumed his duties, Grandjean recalled the ministry's cartographers to Paris to work on France's borders in newly conquered areas; he retired shortly thereafter.[16]

In the meantime the government had decreed (on 1 June 1791) that primary responsibility for mapping boundaries would hereafter belong to the army. No doubt the army had long been jealous of competition from the diplomatic corps. The government may have considered the mapping team in the diplomatic service to be a wasteful duplication of the army's effort. When the government was obliged to make fresh maps of its boundaries in 1814 and again in 1816, army engineers rather than officials of the Ministry of Foreign Affairs did the job. Nominally a diplomat was in charge, but in fact engineers from Foreign Affairs were scarcely involved. The army's engineers consulted older maps, updating some and making new maps when older ones were inaccurate. They also traced all the various borders agreed upon or proposed by Spain and France since about 1500 onto one map. At the end of the Napoleonic Wars, the French were better organized and began to work sooner than the foreign teams assigned to work opposite them. After a burst of activity that lasted until 1818, the pace of work slowed as funds were cut back. Many maps were not completed and certified until the end of the 1820s, and as late as 1832 the Rhine border map was still unfinished.[17]

The history of the mapmaking office of the Ministry of Foreign Affairs in the nineteenth century reflects well on its accomplishments between 1775 and 1793.[18] In the early nineteenth century, the elder son of Barbié du Bocage joined the team when a vacancy occurred; when Chrestien de la Croix retired, he became the titular head and eventually received the title "géographe du Ministère des Affaires Étrangères" and the responsibility of preparing maps to be used in negotiations. When he died in 1843, someone named Darmet took over, but in a cost-cutting move he was forced to work alone until his death in 1852. He in turn was replaced by Buisson, a man who was probably chosen because he would be unlikely to try to expand the geographical service. Whether diplomats were in fact indifferent to geographical knowledge at this time or were simply unwilling or unable to support funding for a geographical office is a topic for further study. Things began to change when Louis Edouard Desbuissons was appointed in 1865. He used his spare time to gain familiarity with the map collection and applied his knowledge during the negotiations following the Franco-Prussian War of 1870–71; apparently he insisted that the road between Belfort and Epinal be retained by the French.[19] Under the Third Republic, imperialism generated more work for a diplomatic geographer, so in 1882 Desbuissons hired assistants. In 1885 he took part in the conference on Africa held in Berlin, and he also participated in negotiations with the Portuguese on African borders between 1885 and 1891. He retired in 1900.

That the geographers employed by the Ministry of Foreign Affairs accomplished far less than the engineers employed by the army or navy can be explained largely by the long lead those units had in acquiring a mapmaking capacity and by the larger numbers of people they employed in that activity. The navy was responsible for colonial affairs, and hence for borders outside Europe; its border maps have been excluded from this study. The mapping functions of army engineers affected the national map survey during and after the Revolution. Soldiers studied border regions in times of peace so as to prepare for war. From 1749 to 1754, Montanel, de Bourcet, and Villaret conducted field surveys of the Haut-Dauphiné at a scale of 1:14,000. The Alpine frontier between France and Piedmont was mapped between 1748 and 1760, largely under de Bourcet's direction. The Comté de Nice was covered in 1763 and 1764, and the coast of Provence was surveyed from 1764 to 1769. Under Jean-Claude Le Michaud d'Arçon (1733–1800), surveys were made along France's eastern borders from the Jura Mountains to northern Alsace between 1779 and 1787 in 313 field survey sheets. D'Arçon himself surveyed France's border on the Rhine in 1786 and 1787. These surveys, at a level of detail much greater than the Cassini map surveys, are among the greatest achievements of French cartography.

Each of these mapping projects took only a few years because working methods were standardized. Fieldwork continued during fair weather from mid-spring to mid-autumn; the rest of the year was devoted to analyzing and copying the previous months' fieldwork. One report enumerated the ideal qualifications for this kind of work:

> An *ingénieur géographe* should know enough geometry to be able to determine geodetically the principal points on the maps he is making. He should be able to make the most accurate topographical maps, being familiar with the representation of all manner of landforms and land uses. He should also be able to depict an expanse of countryside without the use of any instrument, either on foot or from horseback, as field reconnaissance for generals to use in determining the order of march, campsites, battlefields, entrenchments, communications and supply routes, and so forth. Such reconnaissances need to be made as close to the enemy as possible. He should also be able to draw with perfection the structure of all manner of fortifications, offensive and defensive operations, the workings of bridges and field equipment, the countryside, and architecture. He should know how to read and write foreign languages and be able to swim, so that nothing need prevent him from executing urgent and important reconnaissances.[20]

How many of those who served were so qualified? Without an exhaustive study of their careers it is impossible to say. Many fieldworkers were poorly educated and served only one or two years. These cartographic projects required continuous supervision and guidance from officers.

Every spring, commanders gave each team its instructions in great detail. A few months later, officers wrote reports on what had been accomplished, emphasizing information of strategic and tactical value. The maps themselves, together with written texts describing natural resources, economic activities, and topographical features, were to be used by general staff officers in planning fortifications, maneuvers, and war strategies.

In drawing up the instructions for mapping the Dauphiné and Provence in 1775, d'Arçon urged that written memoranda accompanying the maps be brief, since reading such texts was tiring and would deflect attention from general matters to detail. His orders were to make maps that could speak for themselves ("faire parler la carte elle-même"). The more complete and intelligible the maps, the briefer the memoranda, and the better the analysis of topographical features in terms of general principles of warfare.[21] D'Arçon urged his men to synthesize field surveys into finished maps as quickly as possible, pointing out that one quickly forgets things and rarely has the opportunity to go over the same ground twice. He also emphasized the correct spelling of place-names and the unambiguous matching of a place with its name on a map. This was important to the staff officer who would be simultaneously looking at a map and reading a memorandum. D'Arçon also urged his men to note historical or popular traditions unique to given communities, whether as political intelligence or just as "objects of simple curiosity."[22]

In an undated document, d'Arçon developed his views on the proper and improper uses of maps in military service.[23] D'Arçon's main point was that topographical surveys are potentially damaging to military operations if they are used by the general staff to dictate such detailed matters as the order of battle. Certain decisions should be made only by men in the field in a position to take in the situation at a glance and to communicate rapidly, he argued. Maps, d'Arçon believed, are incomplete, no matter how good they are. When they are well made, they create an impression of being comprehensive; the staff may therefore rely upon them too much. Maps are good for showing where the enemy is, the distance between places, the routes from one place to another for different kinds of military units, and the availability of local resources. They should be used to reach strategic decisions involving many different units over a vast area when their movements could be coordinated into a plan. But if too many details are placed on maps, then the staff will spend too much time making tactical decisions that are not its proper responsibility, at the expense of strategy. D'Arçon anticipated that as armies moved faster and covered longer distances, they would need to travel with collections of topographical maps and memoranda. He intuitively understood new ideas of warfare based on mobility and surprise, on control of transportation routes, and on flexibility. It would be ironic, thought d'Arçon, if the perfection of mapping encouraged the decay of strategic thinking.

The border-mapping activities of the army and of the Ministry of Foreign Affairs show that permanent mapmaking and map-collecting functions were taken seriously in the last decades of the Old Regime. Still, these functions were not yet clearly defined in any given unit. Because ministers anxious to cut budgets often attempted to eliminate map-related functions in units or to consolidate units with such functions, persons with cartographic interests were eager to enhance their specializations in order to preserve their bureaucratic autonomy. The use of maps helped to rationalize diplomatic procedures and strategic thinking. But the cartographic activities of the state were not yet themselves rationalized.[24]

On the one hand, cartography nurtured the emergence of a distinctive professional group, trained according to a standardized curriculum, funded by contracts, administered with detailed budgets, and judged according to authoritative criteria of excellence. On the other hand, cartographers in government service had to advance their careers competetively, obtain patronage, and occasionally resort to questionable financial practices. Only the government could coordinate the work and underwrite the costs involved in many of the maps that were needed, but the government was not always willing or able to provide political and financial backing. Only a gifted and determined individual could master these circumstances, but in the 1780s there appeared to be no one active in cartography or geography comparable to Vicq d'Azyr in medicine, Lavoisier in chemistry, or Condorcet in social science, "each envisaging the whole of society from the standpoint of his own speciality (and vice versa)," each attempting to reform a body of knowledge and the institutions through which knowledge is diffused and applied.[25] If maps were used to effect economic, social, and political change, could new maps be made fast and often enough to be valid? How could the expense of mapmaking be justified given the state's financial condition? Was cartography elitist or in the public interest? What could scientific ideas and methods contribute to map-related projects and cartography? These issues, already apparent in the 1770s and 1780s, became important policy matters after 1789.

The Cadaster to 1799

Scientists were unable to prevent—and many actually encouraged—the application of scientific methods and ideas to social, political, and economic issues. The use of maps to foster reform expressed a belief in the compatability between natural laws and social needs. It assumed that man's knowledge of the world and control over his environment would be beneficial. The more reforms were seen to be desirable and feasible, the more scientists were caught up in movements that responded to nonscientific factors, with consequences for their activities that scientists in general, and cartographers in particular, could not always control.[26]

The development of the cadaster (which combined maps and lists to form a land registry) highlights several important aspects of cartography during the Revolution. The cadaster involved the application of cartography to political and economic reform; it was affected by political and economic circumstances; it contributed to the establishment of the metric system; and it shaped the fate of the national map survey beyond 1815.

The few cadasters that were in fact completed during the Old Regime were not part of a national land registry but instead were made for judicial or fiscal purposes according to local needs and traditions. The largest unit involved was only that of the généralité. Reform-minded officials, such as Anne Robert Jacques Turgot when he was intendant of the généralité of Limoges (1761–64), encouraged their use. By the time Turgot had entered his position, some six hundred settlements in the généralité had been surveyed. Turgot recruited Pierre Cornuau, a topographical engineer, to the généralité to be inspector-general of manufactures, and he had a small book by Cornuau on surveying published. (Cornuau stayed in Limoges until 1790.) Surveying continued at a rapid rate; the inspector general of the généralité for the administration of the Ponts et Chaussées estimated that by 1779, three-fourths of the généralité had been surveyed.[27] But there were problems involved with making such cadasters. The surveyor was paid by the community, which was often tardy in collecting for his fee; surveys were often opposed by the local notables, who were afraid of higher assessments; and even the peasants themselves were leery of surveyors for the same reason. Sometimes the teams making the great Cassini map of France were attacked by villagers who confused their work with tax-related surveys. Only in an exceptionally poor area such as the Auvergne did peasants actively petition the king (in 1786) to encourage land surveys, arguing that they were overtaxed and believing that a survey would shift the tax burden to larger landowners.

Few areas were completely surveyed in this system of cadasters. Savoy, for example, had a cadaster made between 1728 and 1731 (the map part being finished in 1737 at a scale of 1:2,400). But Savoy had a different set of legal traditions and also enjoyed good relations with Piedmont, where cadasters were more common.[28] The government initiated a cadaster for Corsica in 1770, but it was not finished until 1793. To verify the triangulation of Corsica undertaken in conjunction with the making of the cadaster, the Academy of Science arranged with the navy and the courts of Turin and Florence to provide assistance with observations between the island and the mainland. Joseph-Bernard Chabert, Joseph Lalande, Jean Bailly, Alexandre Pingré, Pierre-François Méchain, and the Marquis de Condorcet evaluated the results in 1791, the closest that members of the Academy had yet come to work on cadasters. The Corsican operation was the most sophisticated in the Old Regime, with twenty-eight employees, including two translators, fifteen surveyors, two draftsmen, and two statisticians.[29]

The next step in generating a national cadaster was taken in 1763, when Henri Bertin, then controller general, pursuing an order of the previous year to take a census and assess property, wrote to the intendants in the jurisdiction of the Parlement of Paris to ask their advice about making a cadaster. Then in 1778, Dutillet de Villars submitted a prospectus for the establishment of a national cadaster. In 1738 he had offered to make a cadaster of the parish of Montoneau in the Limousin at his own expense, to demonstrate the value and feasibility of cadasters. Orry accepted his offer but paid for the cadaster anyway. On the basis of the Limousin experience with cadasters under Turgot, Dutillet de Villars estimated that 180 towns and parishes could be surveyed at a cost of 29,250 livres (at a rate of four sols, six deniers the arpent). He also suggested using the foot as the standard measuring unit, divided decimally.[30] In 1789 F. N. Babeuf and J. P. Audiffred estimated the cost of surveying at five to eight sols the arpent and argued strongly that the advantages were worth the expense.[31] At least half the cost of a cadaster could be directly accounted for by the cost of making maps. Etienne Munier, who worked in the Limousin as an engineer, published a two-volume treatise on geography in 1779 in which he also discussed the costs and organization of a national cadaster. He proposed hiring unemployed students of architecture, engineering, geography, and drafting. After some specialized training in a school set up for that purpose, they would be paid a salary of 1,500 livres a year. Munier estimated that one person could make a survey of a parish in fifteen months and could assess land values in three. If 24 persons were employed to survey 290 parishes of the élection of Angoulême, it would take eighteen years, at a yearly cost of 36,000 livres, for a total of 648,000 livres, or 2,500 livres per parish. To survey the entire généralité of Limoges would cost 2,592,000 livres and employ 96 professionals for eighteen years. To survey France, 3,072 engineers would be needed, at a cost of 82,944,000 livres. How could the reform of the tax structure generate enough additional wealth and stimulate sufficient productivity to recoup such an investment?

For a national cadaster, a large number of easily trained, semiprofessional, modestly paid surveyors would have to be supervised by an elite group of scientist-administrators. Only centralized control and a uniform measurement standard could ensure uniform results and geodetic knowledge of sufficient accuracy; otherwise, errors in the maps of small units would affect maps of larger units, with a cumulative effect. Cadasters made locally were the work of a generation; yet changes in land use and value over time required revisions. Any cadaster for the nation would have to be easy to revise. During the Revolution a serious attempt was made to solve such problems.

The Revolution began auspiciously enough for cartography with the substitution of departments for the administrative units of the Old Regime. In the eighteenth century, the revision of administrative boundaries had been promoted in the belief that in smaller and more homogeneous political units, informed decision making by consensus might flourish. On 7 September 1789 Abbé Sieyès called for the establishment of a committee to draw up a plan to unify France through redrawing administrative units according to uniform principles. On 29 September 1789 a "comité de constitution" reported to the Assembly on a reorganization of political jurisdictions. A debate erupted between partisans of equal division according to area and of equal division according to population; the former proposed eighty departments about 324 square leagues in area, the latter seventy departments of about 360,000 persons. In addition, people argued about how much the boundaries of the new divisions should deviate from the older ones and from features of the landscape. A "comité de division," supervised by Cassini IV, set to work on the voluminous reports submitted by hundreds of committees across France. The new division of France was presented by Cassini IV in the form of a map on 10 April 1790 (figs. 7 and 8).[32] It was based neither on mathematical principles nor on physical geography, but blended both with local tradition and precedent. The names of the new departments carried references to waterways in sixty-one cases and to mountains in twelve. These references to nature were designed to substitute timeless landmarks for the accretions of history as emblems of common identity and political consciousness.

The revision of administrative boundaries was of course symbolic of the break the Revolution made with the past. By remolding the country according to more rational criteria, including references to nature, the government hoped to cut the knots that had tied up reforms in the Old Regime. Among the many reforms proposed in the early months of the revolutionary era, two related projects—to modernize the measurement system and to establish a cadaster—were conspicuously carto-

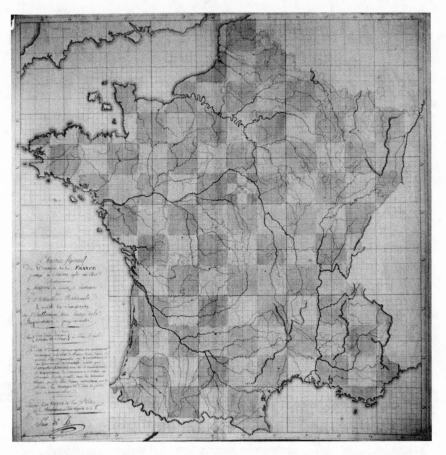

Figure 7   Outline for administrative reorganization of France. The checkerboard grid and the proposed administrative units of equal size were derived from maps by Robert de Hesseln, September 1789. "Chassis figuratif du territoire de la France." Paris, Archives Nationales, NN 50/6.

graphic in nature. But the government enjoyed less success in these matters than in administrative reorganization.

As with the cadaster, proposals for revising systems of measurement had been put forward during the Old Regime. In 1670 Gabriel Mouton, a Lyonnais astronomer, proposed that a universal, international standard based on land measurement be adopted. He sent a copy of his book to Huygens, who in 1673 proposed using the seconds pendulum as a standard. (Picard had made a similar proposal in 1671.)[33] Obvious inadequacies of instrumentation prevented further consideration for decades. In 1766 one attempt was made at harmonizing the units of the Old Regime when Louis XV decreed the toise de Péru to be the national standard and ordered eighty copies of the toise used by La Condamine and Bouguer. La Condamine had earlier suggested taking the length of the seconds pendulum at 45° latitude and subdividing the period of a day decimally. (Since the second is 1/86,400th of a day of twenty-four hours,

Figure 8 First proposed division of france into departments. Signed by Philippe Hennequin, "successor to Robert de Hesseln," this map, prepared for the Assemblée Nationale in September 1789, shows the proposed outlines of departments. "Carte de France divisée selon le plan proposée à l'Assemblée nationale par son comité de constitution." Paris, Archives Nationales, NN 50/7.

decimalization would require reference to a hypothetical day of 100,000 seconds.) Further, he held that the unit should be selected, measured, and adopted as a multinational venture. Apparently this was what Turgot had in mind when in 1775, in consultation with Condorcet, he asked Charles Messier, an astronomer, to travel to Bordeaux (near the forty-fifth parallel, equidistant between the equator and the pole) to make an exact measurement of the length of a seconds pendulum as a standard for other units. Messier never made the trip, however, because his timepiece was broken. Turgot was replaced in office during the time of the repair by Jacques Necker, who considered a reform of weights and measures impractical. Some examples drawn from antiquity were presented to make the point that revisions of systems of measurements to make them uniform and permanent were nothing new;

but Condorcet, for one, felt that neither the tools of classical scholarship nor the measurement systems of antiquity were sufficiently rigorous. He believed the case for reform and for the selection of a new standard could be made solely on an appeal to reason, on the grounds that a system based on nature will bring man's activities into harmony with natural law. Some in England supported this cause, and after the Revolution began Talleyrand urged France and England to adopt a uniform standard as a means of bringing the two nations closer.[34] The hostile attitude of other nations to revolutionary France, however, brought the French to realize that if they tied the introduction of a new standard to the cooperation of others, measurement reform would not come soon.

As early as 27 June 1789, the Academy of Sciences appointed a commission composed of Antoine Lavoisier, Pierre-Simon de Laplace, Mathurin-Jacques Brisson, Mathieu Tillet, and Pierre Le Roy to study weights and measures. Brisson's memoir of 14 April 1790 on the relation of linear units to units of volume and weight has been lost, but either it or a report of 9 February by Claude Antoine Prieur de la Côte d'Or (while still an army engineer) influenced Talleyrand when he wrote the initial metric law, adopted by the Constituent Assembly on 8 May 1790. This law ordained the seconds pendulum as the basis of linear measurement and specified that gravimetric units were to be related to linear units through the weight of water at a given temperature. Cassini IV, a member of a commission on weights and measures of the Constituent Assembly, supported the seconds pendulum. He was of the opinion that a revolution was not a propitious time for geodetic measurements that involved transporting large and delicate instruments about the countryside, an activity that had aroused the hostility and suspicion of people even in more tranquil times. Cassini eventually resigned, to be replaced by Jean-Baptiste Delambre, an advocate of geodetic surveys. In any case, the decisions reached by the Assembly's commission and the Academy's panel may have been academic, insofar as the purpose of the report and law of 1790 was to foreclose the possibility of simply remaining with the traditional units of the Old Regime. To design the new measurement system, the Academy appointed a new commission consisting of Jean Charles de Borda, Condorcet, Laplace, Joseph Louis Lagrange, and Gaspard Monge. Their report was made public on 25 March 1791 and enacted into law the next day. The new commission found theoretical problems with the pendulum, since it made a linear unit dependent upon other variables in nature (time and the force of gravity), and practical problems, since the pendulum is affected by temperature, altitude, and air resistance.[35] Instead, it adopted a geodetic standard.

The reputation of the Academy, already under attack in the 1780s, needed to be enhanced. The scientists who selected a geodetic standard for measurement reform thought this approach would yield accurate results within two years and considered it theoretically superior than any other plan. The geodetic program set forth involved measuring an arc of meridian between Dunkerque and Barcelona, because it was the longest continental line of longitude that did not cross a major mountain range, because it appeared to be international, and because it lay approximately equally north and south of the forty-fifth parallel and therefore should correspond to the mean curvature of the earth.[36] The new unit, or meter, would be 1/10,000,000th of the line of longitude between the pole and the equator, as measured by its span between Dunkerque and Barcelona; it was assumed that this quarter arc of meridian was about 5,000,000 toises, or 10,000,000 meters. (In fact, the meridian quadrant is 10,002,288.3 meters long.)

The relation of a new system of measures to other reforms was obvious to many people. The idea behind the metric system of measures was to make possible the conversion from angular observations of astronomy, the fulcrum of physics, to linear measurements of the earth's surface, which ordinary people use every day, by a simple interchange of units involving rudimentary arithmetic. The simplicity and convertibility of the metric system, achieved largely through decimalization, was fundamentally democratic; insofar as it was easily intelligible, it encouraged independent thinking and reduced the dependency of the mass of people on the most literate and best educated. As Charles Gillispie commented, "The seconds pendulum could never have done that. . . . All the rhetoric about the universality of units taken from the earth itself turns out to have a perfectly sensible foundation."[37] The measurement of the meridian was therefore fundamental, for it translated the determination of the basic linear unit, the meter, into a problem in spherical right triangles, celestial observations converting angular into linear measurements.

Already people were talking about the measurement of the meridian as the basis for the operations of the cadaster.[38] Decrees on tax relief of 23 November 1790 anticipated a new uniform measure; decrees on 4 August and 21 September 1791 stipulated that surveys be made in cases of land disputes; and a decree of 23 September 1791 proclaimed the establishment of a "cadastre général de la France." The Ministère des Contributions Publics turned to the engineers of the Ponts et Chaussées for technical assistance.

The director of the cadaster would have to play many roles: as an administrator, he would establish and supervise the mechanisms for coordinating surveys conducted all over France; as an educator, he would train and certify scores, maybe hundreds, of surveyors and engineers; as a scientist, he would establish surveying on the most accurate measurement system attainable; as an engineer, he would be aware of the implications of surveys for commerce and for the study of economic and social conditions; as a politician, he would have to balance short-term funding problems and the need to show results quickly against the long-term objective of uniformity and accuracy.

Fortunately the right man got the job, Gaspard-François de Prony (1755–1839).[39] Prony graduated from the Ecole Royale des Ponts et Chaussées in 1780 as a protégé of its brilliant director, Jean-Rodolphe Perronet. In 1790 he published the first volume of his *Nouvelle architecture hydraulique*, in which he applied rational mechanics to engineering practice at an elementary level for students. He then became inspector of studies at the Ecole des Ponts et Chaussées. In 1794 he became professor of mechanical analysis at the Ecole Centrale des Travaux Publics (later, the Ecole Polytechnique), serving in that position until 1815. Prony succeeded Antoine de Chézy as director of the Ecole des Ponts et Chaussées in 1798, supervising a reform of the curriculum and a transformation of engineering science to emphasize applied analytical mechanics. He served as director for the cadaster from 1791 until it was terminated in 1799. In the Napoleonic era he served in Italy, where he distinguished himself in such difficult enterprises as the drainage of the Pontine marshes and the development of a new port in the Veneto. After 1805 he served as inspector general of the Ponts et Chaussées and was an active participant on its council, where decisions about public works projects were reached. As director of the cadaster, Prony had the first opportunity of his career to demonstrate the range of his skills. He was of the right age to advance his own career mightily during the Revolution.

Prony submitted a report to his superiors in the Ministry of the Interior on 10 October 1791, less than a month after the cadaster was established by the Assembly.[40] It encompassed all the tasks involved in establishing a cadaster. Either Prony had been thinking about these problems for some time or he immediately grasped the complex nature of his mission. In any case, his memorandum proved to be an accurate description of his activities and of the difficulties he would encounter for the rest of the decade. He began with a critique of the Cassini maps, which he faulted for inaccuracy of topographical detail, small (but nevertheless significant) geodetic errors, and a scale too small to serve usefully in the making of a cadaster. Prony intended to begin the cadaster, therefore, with a thorough revision of the main geodetic triangles of the Cassini map, proceeding to the measurement of at least one secondary triangle per canton. He specified that the length, direction, altitude, latitude, and longitude of each baseline be determined with absolute precision. Uniformity and precision guaranteed the comparability of data across France, thus providing the practical and equitable means of reforming fiscal structures.

The cadaster, as Prony envisioned it, should enable useful knowledge about population, health conditions, commerce, industry, agriculture, and transportation to be collected and analyzed, thus justifying the expense and effort of doing the job right. Indeed, he estimated that the very act of taking surveys would bring to light a great deal of useful information about crops, mineral resources, watersheds, and population. His goal was to combine cadaster, census, and survey of resources into one. To train surveyors and engineers, Prony planned to write a manual on instruments and surveying procedures. New measurement devices would enable surveyors to make their calculations by measuring angles on the land rather than by measuring angles they had drawn on paper based on visual observation and estimates of distance, which were less precise. Aware of the prospect of measurement reform, Prony planned to use the toise de Paris, local units, and an estimated value of the meter simultaneously. He addressed the problem of connecting subsections of surveys to form continuous maps by calling for verification of all measurements and calculations by the central office. Special maps would be needed of areas where several surveys would meet.

To appreciate Prony's approach, compare it with that of Jean Nicolas Buache de la Neuville, Philippe Buache's nephew, who wrote a plan for a cadaster early in the Republic.[41] Buache would have sent each commune a model map, on which various topographical and agricultural features were to be pictured. Each commune would then make a map of its own area, using the model as a guide and indicating the distance from it to neighboring communes. To each map the commune would attach tables indicating the value and size of landholdings, the number of inhabitants according to various age categories, and information about employment, diet, weather, and common illnesses and their treatment. The cadaster as Prony conceived it was a highly centralized, rigorously scientific operation designed to permit the piecing together of maps of local areas and comparisons between them; Buache imagined the cadaster as a decentralized enterprise whose findings would be inherently un-verifiable and difficult to compare. Like d'Anville's attempt at making a diocesan map by sending sketch maps to curates to complete and then assembling them into a finished work, and like Cassini III's suggestion to make a more detailed set of maps of France by piecing together various provincial maps that were based on his geodetic work, Buache's approach made making a cadaster appear easy and inexpensive. To his credit, Prony did not underestimate the money, time, and effort needed to make the cadaster he designed, in order to enhance its appeal to his political masters: "l'exécution ne peut être [que] le fruit d'une constance longue et penible," he wrote.

Prony knew he had no time to spare. Because the Assembly had not provided funds or allocated space in its decree in September, Prony operated the office of the cadaster in the premises he occupied while supervising the construction of a bridge. He hired an assistant director to design a method of classifying land and estimating its value, someone to perform mathematical calculations, a draftsman, and two office boys.[42] He also proceeded to measure the total area of France from the original maps for the division of France into departments. He divided France into zones, whose areas were calculated separately. From these estimates Prony determined the geographical center of France, a point he intended to place at the center of a map of France for the cadaster. It took him nearly a year to estimate the area of France in this way.

Meanwhile Prony launched a most remarkable project in applied mathematics. Inspired by Adam Smith's *Inquiry into the Nature and Causes of the Wealth of Nations*, Prony recruited hundreds of men who knew only the elementary rules of arithmetic. Applying the concept of division of labor, he set them to work making the calculations for new logarithmic and trigonometric tables incorporating decimal divisions. The repeating circle designed by Borda had been converted to decimal divisions from the traditional duodecimal division of degrees in anticipation of the adoption of decimal units for angles. This never happened, but had decimal units been adopted for angles, existing reference tables would have been impossible to use. Prony's new tables were designed to allow the metric measurement team and cadaster surveyors to compute measurements decimally.[43]

And Prony went ahead and estimated the length of the meter. Knowing that one definition of the meter under consideration was 1/10,000,000th of an arc of meridian from Dunkerque to Barcelona, he deduced the length of the new unit with sufficient precision to proceed with the cadaster; his estimate that a meter would be 3 feet, 11 and 481/1,000 inches long was 44/1,000 too much. By 1793 Prony had accomplished enough to submit his plans to the scrutiny of a committee of the Academy (composed of Lagrange, Delambre, Borda, and Laplace) and to the Assembly for their approval. Laplace drafted the report, in which he proposed the terms meter, decimeter, centimeter, and millimeter.

Prony's strategy was farsighted. Ideally, work on local surveys ought to have followed, not preceded, geodetic mapping of large areas, lest insignificant distortions accumulate, making congruence among several local maps impossible. For the same reason, work on the metric system ought to be complete before work on the cadaster began. Prony knew that Jean-Baptiste Delambre (1749–1822) and Pierre-François Méchain (1744–1804), former assistants of Laplace, were preparing to measure the meridian between Dunkerque and Barcelona in 1793. Given their expertise, and taking into account the rate of progress on previous geodetic surveys, Prony estimated that they would need a year for fieldwork and a year for analysis. He prepared, therefore, to launch fieldwork for the cadaster in 1795.

Prony proposed dividing the administration of the cadaster into three sections—geodesy, survey, and land use and valuation. Each would progress at its own rate and could advance independent of the others. The map scale was specified at 1:2,000 for individual surveys, at 1:5,000 for maps of the distribution of land uses in a commune, and at 1:20,000 for geodetic surveys. Taken together, tens of thousands of sheets of paper would be needed to display the cadaster surveys for the entire country.[44]

Anticipating that geodetic fieldwork and local surveys would begin simultaneously in 1795, Prony created the Ecole de Géographie (also called the Ecole Nationale de Géodésie Pratique), apparently *before* the major reforms that created the other "grandes écoles" and independent of them.[45] (Writing in the Old Regime, Munier had conceived of a school for specialized training in surveying.) Prony staffed the school with five instructors: one for land-value assessment, one for field observation and instrumentation, one for calculation of land area, and two for drafting and graphic arts. The school was necessary to ensure uniformity and accuracy of results. Prony planned the school to provide classroom instruction to two hundred students in to- to three-month periods, thus graduating perhaps six hundred or more surveyors in a year—enough to send five or six to each department. (In addition, the school might supply other government offices with geographers and surveyors.) Once in the field, geographers working for the cadaster were to form two teams, one making geodetic measurements, the other making detailed surveys of properties. In other words, Prony intended the school to train enough surveyors so that the cadaster could be launched as soon as Méchain and Delambre completed their surveys for the metric system of measurement. The school was the administrative, pedagogic equivalent of

Prony's logarithmic tables and was as daring in design: both were intended to compress tasks normally requiring a generation into one or two years.

But things did not go according to plan. The practical scientific work involved in determining the new units of measure was unexpectedly difficult. To measure 1/10,000,000th of an arc of meridian, Delambre and Méchain had to determine the exact difference in longitude between Dunkerque and Barcelona and make precise astronomical readings of their latitude; they also had to check the triangulation grid with new observations and calculations and obtain measurements to use in future geodetic work.[46] The mid-1790s were not a good time to travel with a baggage train of large and delicate instruments. Méchain was detained as he left Paris on the grounds that his scientific work was a cover for a counterrevolutionary conspiracy. While convalescing from an accident when examining a hydraulic pump in early 1793, Méchain was also arrested in Spain. In the year that ensued before his release, he went over his calculations and noticed a discrepancy of three seconds in the latitude for Barcelona he had previously communicated to Paris. Unable to perform new observations, he kept his error secret from everyone. His behavior in France after his return in 1795 mystified those of his friends who had survived the Terror, for his anguish over such a small error became a neurotic obsession. From January 1794 to April 1795, the Committee of Public Safety suspended reform of measures anyway. In 1795, General Etienne Nicolas de Calon, director general of the War Department's geographical section, appointed Delambre and Méchain as astronomers, and in 1796 fieldwork was resumed. Much of the work previously undertaken had to be repeated because many church spires, commonly used as fixed points in a triangulation grid, had been torn down in the interim. The two scientists finally joined the separate parts of their grid in September 1798.[47] In anticipation, Talleyrand had already convened an international conference to consider the adoption of the new metric system.

Let us return to Prony in Paris in 1793, when the difficulties confronting Delambre and Méchain began to cause concern among scientists. In 1793, and as far as can be determined for at least the next three years, the geodetic section of the cadaster concentrated on other matters. Prony's tables converting the division of the quadrant from duodecimal to decimal units were prepared for publication by 1794; a press run of six thousand copies was anticipated. (The tables were never printed.) The geodetic section also prepared a map of France divided into departments, based upon and completing research begun in the 1780s and continued by Prony estimating the area and population of each department and average population density.[48] An annotated bibliography of printed works with special attention to the economic activities of villages and farms was launched (but probably never completed). Other projects included a historical treatment of population migration across the globe and making maps of the earth at different epochs in geological and anthropological development; how much progress was made on these projects is impossible to determine.

To fill the time until field surveys could begin, Prony added to the cadaster's responsibilities the task of keeping track of the condition of all transportation routes by land and water, including all public works under repair or proposed for construction. To this end he detailed four draftsmen to make a map of France including major topographical features at a scale of 1:500,000. Prony anticipated relating all cadaster maps to it, using this map to synthesize information about how economic goods circulate across France. But even these tasks, which involved routine work in an efficient office under Prony's supervision, could not go forward smoothly. The more comprehensive the scope of the cadaster, the more other government offices turned to it for assistance in geographic and cartographic matters. Cadaster draftsmen were frequently asked to make maps for the army or for other government offices. The cadaster was organized at a time when the government's attempt to establish a national map archive (described in the next section) created chaos. The cadaster became indispensable by virtue of its order.

Inflation and intense budgetary pressures further complicated Prony's plans. The scale of his school was reduced. In 1796 it was transformed from a large institution recruiting from a broad social base into an elite school for no more than fifty graduates of the Ecole Polytechnique.[49] Called the Ecole des Ingénieurs Géographes, it was further modified by being divided into two parts, one for geography, the other for ballooning (an activity of military value, based at Meudon). Students were supposed to spend one year studying each subject. Since neither geography nor ballooning offered many positions, few students applied. Only twenty out of a possible fifty places were usually filled.

In 1796 the total budget of the cadaster was about 144,000 francs, of which 18,240 francs were spent on supplies. Of the balance, the school received 31,644 francs, mostly for salaries for the instructors and stipends for the students. Prony received a salary of 8,000 francs, a section chief 6,000 francs, a deputy section chief 4,000 francs, a calculator 3,150 francs, a registrar 2,000 francs, a draftsman either 2,500 or 2,000 francs, and office staff between 1,800 and 800 francs. Each of the three geographers and scientists employed by the cadaster in Paris received 3,000 francs. At the same time, the Ministry of the Interior was spending 200,000 francs in stipends to encourage writers, nearly 270,000 francs to support the Muséum de l'Histoire Naturelle, and about 400,000 francs on national festivals. Measurement of the meridian had originally been expected to last two years and to cost 160,000 francs; it cost 360,000 francs by 1794 and an additional 100,000 francs in 1796, considerable but not exorbitant sums.

By 1799, when measurement reform was complete, and when in theory the real work of the cadaster could have begun, budgetary pressures were too great. The government reduced total funding for the school and the cadaster to 100,000 francs and projected their elimination from the budget for the next year.[50] Given the slow rate of progress to date, the government believed the money could be used on projects promising quicker results or more immediate benefit to the nation. Prony complained, not without reason, that he and his staff had been asked to do too much and had been underfunded for too long. To close the program now, Prony argued, would mean wasting all that had been invested and would lead to the dispersal of trained professionals who might not be recruited again later. Thinking of the future, Prony urged that whenever the cadaster was resumed, it retain the principles on which he operated: pursuit of perfection in large-scale and small-scale work, comprehensive treatment of the entire nation and of its human and natural resources, and centralized administrative control.

There are huge gaps in the chronological continuity of the documents relating to the administration of the cadaster in the 1790s, but enough detailed information has survived to compose a picture of the entire enterprise. Clearly, Prony had reason to believe the cadaster, on which the reform of France's fiscal system and the revitalization of agriculture and commerce depended, was underfunded. Proper funding for the cadaster required far greater resources in 1793–96, when so many people needed to be recruited and trained and so much equipment manufactured or purchased. Even if Méchain and Delambre had been able to complete their work by 1795, major funding probably would not have been forthcoming in time to launch nationwide field surveys in 1795 or 1796. Prony's operational plan should not be faulted for suffering from the effects of war and inflation, circumstances clearly out of his control. He had designed a plan that promised scientifically accurate and speedy results. Given the state of geodetic knowledge and technical training, it may well have been the best plan possible, although the sums involved were substantially greater than the government was willing or able to spend. The cadaster of the 1790s does not deserve to be ignored simply because so little was actually accomplished. The record for the 1790s, taking into account all the reasons for delay, was nonetheless impressive. More had been accomplished in this tumultuous decade than in any more tranquil decade of the Old Regime.

The Cadaster after 1799 and the Centralization of Cartography

Part of the problem affecting the cadaster in the 1790s was the government's inability to decide what kinds of mapmaking projects were most deserving of its support. As a result, sponsors or administrators of cartographic projects had to compete against each other as well as against all the other claimants for funds in a situation that may have made some reminisce about the days of ministerial partronage under the Old Regime.[51] Since the government was subjected to claims for funds for cartographic projects involving the making of new maps and the conservation of existing ones, it needed a policy to guide it. But its decisions were often reached in response to pressures from and in consultation with various public officials, some of whom had at best a vague knowledge of maps, so it received information that reflected disagreements among mapmakers and map users themselves—disagreements that the circumstances of the Revolution only exacerbated. Was geography a cumulative, encyclopedic discipline in which almost any map, whatever its defects, had something to contribute? Or was it to be constructed from the ground up, with entirely new, more scientifically rigorous and more accurate maps as the basis for ambitious syntheses of knowledge? Which approach could best meet the needs of the state? While each had its advocates who used rational arguments, ultimately decisions were based as much on expediency as on logic. To the extent that politics dictated the conditions under which maps were made, collected, analyzed, and used, both approaches received partial but inadequate support in the 1790s, with effects on cartography that were felt well beyond 1799.

In the 1790s both approaches had the effect of strengthening the centralizing tendency of the French state. Prony's reason for a highly centralized operation was that it alone could guarantee uniform procedures and uniform results. But his was not the only effort to centralize cartography in revolutionary France.

The problem of map collections was taken up in 1793 by the temporary commission on the arts established by the Convention. One of the commission's thirteen sections was staffed by Gaspard Monge and Jean Nicolas Buache de la Neuville, who were responsible for designing a policy for collecting materials relevant to naval affairs and geography. They would have omitted nothing. To be included were celestial and terrestial globes, as well as armillary spheres and planetariums over one foot in diameter; atlases and map collections by the best authorities; all available foreign maps and maps of foreign nations; manuscript maps of France's frontiers and coasts, of its colonies and neighbors, and of places frequently visited by its merchants; all manuscript maps of harbors and anchorages, of fortified cities and forts, of battle-grounds, and of projects for roads, canals, ports, and other public works; travel books, ships' logs and journals; models of ships and of machinery used in ports and aboard ships and engraved copper plates illustrating these things; and finally, navigational instruments. Such a collection by itself would have been an enormous venture at any time, let alone in 1793. Perhaps Monge and Buache thought that no one could object to such a reasonable and desirable objective. Compared with what happened next, their proposal almost appears modest.

The most uncompromising effort to centralize cartography involved the establishment of a French national map archive.[52] By decree, on 20 prairial of the year II (8 June 1794), a map office was established under the direction of the Commission des Travaux Publics to collect and classify all maps and geographical memoirs and publications gathered by government officials. This office would replace and expand the War Department's map office, which had survived from 1763 to 1791 without major changes. Eventually the Committee of Public Safety ordered all agencies and departments to deposit their maps with the office, which was also authorized to receive maps from private citizens. Publication of all geographic and cartographic materials was forbidden without the express consent of the office, thus allowing the government to centralize and monopolize cartography.

Some government departments were unwilling to cooperate, for their staff knew only too well how difficult it had been to form their collections. The diplomatic corps attempted to retain in its possession maps annexed to border treaties, maps attached to political memoranda, and maps and atlases used daily by division heads in the conduct of foreign affairs, but the Commission des Relations Extérieurs was nonetheless ordered to submit a list of maps considered indispensable in diplomacy to the Commission des Travaux Publics for that body's approval. Hostility was not the most effective weapon against the archive. Intentionally or inadvertently, by cooperating in the creation of a national map center, government officials in various ministries ensured its failure. The center was overwhelmed and ill equipped. Since its facilities were also used by other government officials, many maps were lost, damaged, or stolen. Soon the order came to return military and naval maps to the War Department's map section, fortifications materials to the Council on Fortifications, maps related to public works to the Ministry of Public Works, and diplomatic maps to the diplomatic corps. But the original decrees ordering maps to be deposited remained in effect, so more maps continued to arrive at the same time. Prony was obliged to second some of his draftsmen to make maps for various government offices at this point, as has been noted. Maps were mixed up. Lafayette never received the manuscript maps made in Virginia by Louis Capitaine, then his aide-de-camp; they had been sent to Louis-Lazare Hoche for the army of the Sambre-et-Meuse. In these conditions, map-related activities of many government offices, such as the registry of land titles involving ecclesiastical and aristocratic properties, were impeded.

After the map center was disbanded late in 1794, representatives of several ministries met to consider consolidating their map holdings in the Louvre in the form of a "musée de géographie et de topographie militaire et d'hydrographie," to be staffed by the individuals who had custodial care of maps in their respective ministries. In this plan, the identity of each map collection would be preserved within a large administrative framework, thus avoiding the problems that had so rapidly overwhelmed the national map center. It was proposed to make a catalog with entries for each map specifying its subject matter, sources, authors, and such technical matters as its projection and accuracy. In addition, the museum staff was to make new maps for civilian and military purposes and to provide instruction in geography and topography. The professional staff of the Musée de Géographie was supposed to be composed of topographical and hydrographic specialists. In time of peace, the former would make maps for the army, maps of land frontiers, maps and charts of land elevation, geological maps, and inventories about trade, natural history, and weather; the hydrographers would systematically survey inland waterways and the oceans. In time of war, topographers would perform military reconnaissance and train staff officers, while hydrographers would serve with fleets on battle stations. In addition, the institution would employ two astronomers (to finish the Dunkerque-Barcelona geodetic survey); a mathematician; specialists in ancient, modern, and physical geography and in naval and military history; a mathematics professor, a translator, an archivist, and a bibliographer; and clerks and engravers. Centralization on this model would have allowed state agencies to maintain their own cartographic units while subordinating them to a centralizing, coordinating unit. This enterprise with a proposed staff of one hundred was terminated as too costly, and its component parts were reallocated to various government departments.

Calon, whose idea this had been, committed the mistake of trying to accomplish too much too fast at a time of rapid political change and with inadequate funding.

The War Department's map division (under Calon) merged with the navy's hydrographic office and the topographical section of the Committee of Public Safety. (Since 1794, the Committee had tried to keep track each day of the movements of armies on maps, even by using pins to identify the strength and composition of units.)[53] A year later the topographical section was disbanded, and the war and navy collections were reconstituted as separate units.

The spirit that animated these attempts at building a national map center was not quenched by failure. It survived in the courses which Buache de la Neuville and Edme Mentelle gave at the Ecole Normale in 1795, and in the commission Mentelle received in that year from the Comité d'Instruction Publique to edit an elementary geography text.[54] Mentelle's book, *Géographie enseigneé par une méthode nouvelle; ou, Application de la synthèse à l'étude de la géographie, ouvrage destiné aux écoles primaires*, published in 1796, promised more than it delivered. Nicolas Demarest had already published a *Dictionnaire de géographie physique* based on the view that general principles and laws could be extrapolated inductively from observation and fact, including the study of information on maps. Mentelle also adopted induction as his pedagogic technique. He would have the student begin with his own local area and then expand his knowledge, by including first all the departments within fifty leagues of his home and then those within a radius of seventy-five leagues, and so on until he reached the national border. This method, whatever its pedagogic value, was neither new nor a synthesis; it presumed the existence of principles of geography but did not spell them out. Geography still labored under the weight of its past, which included maps of greatly varying accuracy, an impressive but limited and excessively descriptive vocabulary, and too few accurate instruments. Mentelle's book merely reflected the lack of critical thinking and the untempered ambitions that had dominated the consolidation of map collections.

When Napoleon assumed power, therefore, the state's ability to control vast cartographic collections and ambitious mapping projects was seriously in doubt. Thus when Napoleon reconsidered the problem of the cadaster, he wanted to use administrative procedures and cartographic techniques that differed from those of the 1790s. To simplify and accelerate the implementation of the cadaster, Napoleon allowed for a greater measure of decentralization and for lower scientific standards than had Prony and the legislative assemblies of the 1790s. Successful attempts to abolish the metric system during Napoleon's reign and afterward underscore the compromises made in the cadaster's scientific integrity.

On 12 brumaire year XI (3 November 1802) it was decreed that cadastral surveys on the level of the commune be undertaken.[55] No fewer than two or more than eight communes for each sous-préfecture were to be selected for surveying by lottery in Paris. Each survey was to be mapped at a scale of 1:5,000 on a single sheet by a surveyor appointed by the prefect, with expenses underwritten by the local community. Originally, four copies were to be made. The scale was too small to show individual landholdings; the maps were detailed enough to show only land use. But without greater precision, verification was not possible, and much of the information was depicted inaccurately to reduce a community's tax burden. The maps had to be used with two sets of books, one recording the land use of each property unit, the other recording all the properties of each landholder. The flaws in this procedure soon became apparent.

In 1807 the procedures for the cadaster were revised. Instead of treating the land of a commune as a unit on a scale of 1:5,000, the surveyors were to map individual landholdings, or *parcelles*, at a scale of 1:2,500, and sometimes at a scale of 1:1,250. It was estimated that at least 100,000,000 landholdings existed. The Ministry of Finance established a commission headed by Delambre to codify procedures, and in 1811, a *Recueil méthodique des lois, décrets, reglements, instructions et décisions sur le cadastre en France* was published. This volume strongly influenced procedures to survey landholdings throughout the empire, even after 1815. Fieldwork began on 3,200 communes in 1808 and on an additional 2,000 more in 1809. By 1814, 9,000 communes had been included, covering 12 million hectares and 37 million parcels. From the perspective of the cadaster as planned in the 1790s, the cadaster of 1807 appeared to place great emphasis on simplicity and efficiency. It was a highly decentralized operation, in contrast to Prony's. Prony had valued accuracy and uniformity above all, and he believed that only a centralized administrative structure could guarantee them. As organized in 1807, the cadaster amounted to nothing more than a land survey, whose accuracy was guaranteed only by the individual surveyors. It was divorced from any large-scale map of France or national geodetic grid. The assumption was that the errors of any one survey would be compensated for by the errors in others and would be so minute as to have no effect on the relation between surveys in any two places.

Here we have the paradox of so highly centralized a state as France, with so long a tradition of state patronage of science, choosing a decentralized and unscientific approach to the cadaster as an expedient means of reforming the tax system, itself a principal support of the centralized state! No doubt to Napoleon and his collaborators the procedures of 1807 seemed necessary if the cadaster was ever to be established. In fact, they do not appear to have been farsighted. Although the organization of the cadaster of 1807 appeared to be much simpler scientifically and administratively than Prony's cadaster, coverage of France was accomplished as slowly as if sounder procedures had been followed. Worse, the cadastral survey could not be revised to reflect changes in land use or ownership. The original cadaster of the first half of the nineteenth century is valued by historians precisely because it reflects the condition of the land at a given time, but this is so because it was impossible to change it. No doubt for fiscal purposes a permanent, invariable record has much to recommend it. But it had always been the intention of reformers that fiscal reform would encourage a more productive and rational use of land, and hence successive changes in the landscape that the survey would have to record.

The cadaster of 1807 inverted the procedures Prony had adopted when he decided to coordinate land surveys in reference to a geodetically accurate map of France. It was only a matter of time before someone, following the logic of the cadaster of

1807, had the idea that a map of France could be put together by assembling and reducing all the land surveys of the cadaster. In 1814 Laprade, *ingénieur géographe*, proposed this very thing to the Ministry of Finance. His idea was to compose the local surveys into maps of cantons divided into communes, at a scale of 1:20,000, with an average of ten geodetic triangles (the third, or lowest, level of triangles). These in turn could be reduced by one-third to compose maps of arrondissements divided into cantons, and so on. Laprade thought this could be done simply by sending copies of surveys on tracing paper to Paris, where trained draftsmen could perform the necessary reductions. He argued that this work ought to be centralized in Paris, not out of concern for accuracy or uniformity, but because it could be done there more cheaply—by half, he estimated—than by specialists in the several provinces working at piece rate. The cost, he figured, would be 4.5 centimes per hectare, or a total of 2,340,000 francs, simply to transform the cadaster into a map of France, or about 30,000 francs per department.

Thinking this might be a cheap way of getting a map of France that could meet the needs of civil and military engineers, who were dissatisfied with the Cassini maps for being of the wrong scale for their purposes and for omitting topographical features, the government turned Laprade's proposal over to a commission of engineers at the Ponts et Chaussées, under the chairmanship of Prony.[56] Prony approved the principle of combining a general map of France and the cadaster as fulfilling the intention of the Constituent Assembly in 1791. Admitting he knew nothing about how the cadaster had been organized in 1802 and 1807 (perhaps he had not been kept informed because he might have objected to the principles and methods adopted), Prony raised some fundamental questions about the geodetic accuracy of cadastral surveys. He observed that the geodetic points in the surveys had to be related to the national geodetic matrix if the surveys and a national map were to be congruent. Unless this could be assured, Prony wrote, surveys that may be accurate at their original scales may be in error when fitted together with other surveys. Not surprisingly, Prony suggested reversing the order Laprade proposed, starting with a geodetically valid map of France instead of with a cadastral survey. Prony believed that perfection in matters of accuracy was desirable and attainable. Arguing that accurate maps give rise to fewer problems in the long run, Prony presented the pursuit of accuracy as an economical and sensible investment. In his work for the cadaster in the 1790s, he had tried to compress the time needed to train a professional staff and to undertake fieldwork, so that extra care and proper methods would not mean greater costs or delays. Unfortunately, the government perceived centralized administrative control over mapping to be too costly; ironically, the economies the government thought it was obtaining were illusory.

Cassini IV, writing about the cadaster and national mapping survey proposals of 1817 in his unpublished memoir, also dealt with the question of accuracy. More cynical than Prony about government, Cassini believed that institutional, bureaucratic factors would defeat attempts to ensure uniformity and high standards. He would have contented himself with limits to accuracy set by need and utility, calling accuracy beyond this "superfluity, vain luxury and charlatanry." (Cassini also foresaw that the cadaster would take too long to make and that delay would have the effect of perpetuating inequality and injustice in the assessment of taxes.) Prony and Cassini agreed that unless professional scientists able to envision complex mapping projects in all their detail were in charge, such projects would not serve the best interests either of the state or of science. They had reason to be concerned about the ability of bureaucrats to manage the task. An astronomer who traveled around France in 1818 to evaluate cadaster fieldworkers concluded that more than half lacked the training or ability to perform even the simplest of their tasks competently.[57] Joining the cadaster and the national survey together would not be easy.

Whatever Prony—or Cassini IV, for that matter—thought of Laprade's proposal ultimately carried little weight, because the army engineers made a similar proposal, independent of Laprade's initiative, that affected more decisively the future of both the national mapping survey and the cadaster. The army's topographical engineers were in an uncomfortable position in the years immediately following the fall of Napoleon. Heirs to the Cassini survey (the publication of which was resumed in 1815, having been suppressed by Napoleon) and of Napoleon's numerous mapping ventures, and custodians of all the maps made or seized in the course of his conquests, the engineers were unable to secure the funding needed to continue work in progress or to prevent the allies from repossessing maps the French had captured from them. Having executed impressive geodetic operations throughout the empire, these engineers knew they could do the work of the cadaster staff much better than it could. And they knew that a new map of France with revisions of Cassini's geodetic work needed to be done anyway. Corrections to add administrative changes and improvements to highways, undertaken routinely since 1793, would never remove the geodetic and topographical inadequacies of the Cassini maps; and the copperplates could be printed only a certain number of times (three thousand) before deteriorating. Plans for a new survey were elaborated in 1812 and again in 1814. In 1816, Simon-Pierre Brossier, assistant director of the Dépôt de la Guerre (the army's cartographic repository), and Maxime-Auguste Denaix, a member of the general staff, conceived a most detailed project involving fabrication of a map equally suitable for use as a cadaster or as a national survey, because it would be made at a very large scale and could be reduced as needed.[58] Their intention was to concentrate responsibility for this enterprise in the army. They believed, as did Prony and Cassini IV, that only centralized control could ensure uniform results and a high standard of accuracy. The army would supervise the third or smallest network of triangles for the cadaster. To this would be added information culled from the collections of the fortifications engineers, the forestry administration, the navy, and the civil and mining engineers. In the opinion of these officers, most of the materials for the new map already existed. What was needed was an agency to coordinate and harmonize these materials, to subordinate a disparate and heterogeneous body of documents to uniform concepts of geography and rigorous standards of geodesy. Would the army's Dépôt de la Guerre succeed where several experiments in the 1790s at establishing a national map center had failed? Events soon disproved such optimism.

In 1817 orders were issued to dissolve the Dépôt de la Guerre as a cost-cutting measure, the staff being redistributed among other units. The staff of the Dépôt, maneuvering to modify these orders, exploited the government's indecision over what to do about the cadaster–national survey proposals. Laplace spoke out in the Chambre des Pairs in favor of combining the two mapping operations if the cadaster could be established once and for all on a rigorous geodetic basis; he believed, mistakenly, that the part of the cadaster already finished could still be used in a national mapping survey without incurring additional expense. Laplace then found himself at the head of a royal commission charged with reaching a decision.[59] The commission quickly agreed to recommend adoption of the army's proposal. But implementation of the proposal was frustrated by several factors that the commission was powerless to affect: uncertainty over funding and staffing levels; lack of agreement among specialists about scales, mapping methods, topographic details to be included, engraving styles and symbols; and the unwillingness of various bureaucratic units, principally the cadaster operation, to share information. There may well have been sound scientific reasons introduced to justify some of the positions taken, but it is nevertheless likely that these negotiations were manipulated by the army to widen its sphere of cartographic influence as much as possible and by the cadaster to preserve its autonomy. The commission met for the twenty-first and last time in 1826. By that time, the army and the Ministry of Finance had taken steps independent of but parallel to each other, the former to start the national mapping survey, the latter to continue the cadaster. As a result, the cadaster continued without a secure geodetic foundation. Even though the army eventually received access to cadaster materials, cadaster maps were so uneven in quality and so defective in geodetic matters that they had limited utility at best for the army's own work. Because the army had to undertake much of the geodetic work appropriate for a cadaster, the final cost of both mapping operations, conducted separately, was probably greater than if they had been conducted as complementary aspects of the same enterprise.

If undertaking a national map survey was part of a strategy to preserve the Dépôt de la Guerre, it was largely successful. The Dépôt was revived in 1822, with a special section for the survey. In 1832 this service was assimilated into the general staff ("l'état-major"). Beginning with a new geodetic grid made between 1818 and 1827, the army proceeded to map France at a scale of 1:80,000; this is the famous "carte de l'état-major." The second and third order of triangles were completed by 1845. Landform details were recorded at a scale of 1:40,000 and were completed in 1866. It had taken the Cassini team from 1750 to 1786 to complete the field surveys for their map of France; forty-eight years were needed to make the "carte de l'état-major." The cost, excluding the salaries of the military personnel involved, amounted to some 12,000,000 francs. Since the legislature never allocated separate funds for this map, the army was obliged to finance it out of its regular account. By the time it was finished it had to be revised, first to take account of the transformation of France's urban and transportation infrastructure that was underway, and second to correct errors owing to cost-cutting measures.

As for the cadaster, between 1814 and 1821 only an additional three hundred to four hundred communes a year were added. The rhythm of work accelerated in 1821, following a reorganization of the cadaster that transferred primary responsibility for its operations to the department and the commune. By 1840 two-thirds of France had been surveyed; the rest was covered in the next decade. The cost between 1807 and 1850 amounted to 160,000,000 francs, or about 3 francs per hectare.

It can be said that the definitive incorporation of the national mapping survey and the cadaster into the working of the French state represented a failure, not a success, for the attempt to bring science and statecraft closer together. Charles Gillispie, in drawing to a close his study of science and polity in Old Regime France, wrote that the role of science "was to provide the monarchy with the services and knowledge of experts and in return to draw advantages from the state for the furthering of science."[60] "From science, all the statesmen and politicians want are instrumentalities, . . . weapons, techniques, information, communications and so on. . . . [Scientists] have not wished to be politicized. They have wanted support, in the obvious form of funds, but also . . . for . . . professional status."[61] Yet here we have a startling example of how scientists can lose as well as gain influence in public policymaking. The decisions affecting cartography in 1807 and in 1817 were not based upon scientific criteria and contributed nothing to the welfare of science. Ironically, these decisions did not produce important benefits for the state either. Given this turn of affairs in France, how could the state promote and nurture innovations in cartography needed when the industrializing process generated unprecedented social, economic, and political problems?

From the example of the cadaster after 1799 we learn that in cartography, as in so many other matters, the Revolution had not brought about profound and enduring changes in the working of the state. Because cartography was so conspicuously involved in and dependent upon proposals for economic change and development, its welfare became hostage to debates over how much change and development were desirable, who would control those processes, who would pay for them, and who would benefit. The Revolution generated a variety of approaches to cartography, but it failed to institutionalize and perpetuate a viable, comprehensive, scientifically valid framework for mapping. In the Old Regime and the Restoration alike, various state agencies sponsored their own mapping ventures. Each group had its own standards for what constituted a good or useful map; each also had vested interest in preserving the autonomy of its own mapping operations. As long as the government did not articulate a coherent policy embracing the cartographic activities of many administrative units, the bureaucratic structure of the French state effectively depressed incentives for interservice cooperation. Yet organized science, especially after the Academy was abolished in 1793, was in no position to advance cartography outside the state bureaucracy and independent of state funding.

Prony had warned the government that the decision about Laprade's proposal was of great significance. He pointed out that France had brought pure and applied mathemathics near to perfection in the late eighteenth century. Other countries, emulating France, had made enormous progress and were ready to take the initiative away from the French. Prony believed that the French government wanted the benefits of science without investing in the patient, costly, and intellectually demanding work that was necessary before those benefits could be reaped. In arguing that the government should invest first in a geodetically accurate map of France, Prony held that scientific criteria were decisive; but political and economic conditions were no more favorable to Prony's methods during the Restoration than they had been in 1799. In France, the separation of science from the cadaster and the national mapping survey had become a political and economic fact, and with that separation ended one part of the late eighteenth-century reformist plan to ground the economic and social development of France in the rigorous scientific study and measurement of nature.

Three

Maps of the Seas

Theory and Practice

The chart of the Gulf Stream by Benjamin Franklin and Timothy Folger illustrates the relation between practical and theoretical concerns that will dominate my discussion of maps of the seas and of mountains. The Gulf Stream flows northward from Florida at a speed of three or four knots. The oldest cartographic representation of the Gulf Stream and of other major ocean currents is believed to be the work of Athanasius Kircher in 1665. According to Lloyd Brown, another map of ocean currents by Eberhard Werner Happel, published in 1685, "looks like nothing more than a scribbling of an impressionistic artist on the loose."[1] Their maps, which showed currents as lines on the otherwise empty surface of the ocean, hint at the enormous imaginative effort that went into representing such invisible natural phenomena of interest to navigators as winds, currents, and magnetic variation. Progress in seamanship led to an accumulation of data that cartographers were not prepared to handle. Benjamin Franklin became involved in mapping the Gulf Stream in 1768 when the authorities in England and America became increasingly frustrated by delays in transatlantic communications and anxious to reduce them.

Franklin was the North American deputy of the British postmaster general. He identified the Gulf Stream as the cause of delays in westbound crossings, and with the help of a Nantucket sea captain named Timothy Folger (the two men were cousins) he supplied London with an annotated chart of the current. The Nantucket mariners knew about the Gulf Stream because they hunted whales in its warmer waters; English captains encountered the Gulf Stream when, for fear they might make landfall too far north, they set their course south, crossed the current, and were pushed farther eastward. Folger marked the points where the Gulf Stream is narrowest and strongest, as well as its breadth, course, and speed more generally from Florida to Canada, "so that ships bound from the Banks of Newfoundland to New York may avoid the said stream, and yet be free of danger from the Banks and

Shoals."[2] Franklin suggested that Folger's remarks and charts be published or otherwise made available to the captains of the mail packets. But the light of truth was weaker than Franklin supposed. As Folger commented, he and others had tried to tell British officers about the Gulf Stream but had been rebuffed. Folger implies that the English were too proud to take advice from colonials. Lloyd Brown has suggested that Franklin's assertive personality and political opinions made him unpopular, causing British authorities to turn to another source for information. William Gerard De Brahm, surveyor general for the Southern District of the American colonies since 1764, was called back to England in 1771 to explain certain irregularities in his accounts. He charted the Gulf Stream during that voyage and published the chart in *The Atlantic Pilot* in 1772.[3]

Further inquiry into the reasons for the neglect of the Franklin-Folger chart by the persons who were supposed to benefit from it has been impeded by the absence of copies in libraries. Franklin mentioned in a letter in 1785 that the British postal authorities had commissioned John Mount and Thomas Page to engrave his chart, and that it had been published by Georges Louis Le Rouge in France[4] (fig. 9). Now that three copies have been located, it is possible to speculate further on why British navigators ignored the information Franklin thought so valuable.[5] A close look at the map suggests that the British had good reason to use Franklin's information with circumspection. The map is a composite of four subcharts joined together along 16° N and 32.5°W. The geographical content of the coastline and the projection had been evaluated by no less an authority than Edmond Halley when the map was first printed, sometime between 1700 and 1720. Its geographical content was therefore dangerously out of date by 1770, when the Franklin-Folger information was *superimposed* on it. Perhaps the postal authorities were responsible for this mistake; perhaps Franklin and Folger had simply used the older Mount and Page chart as the backdrop, so to speak, on which they drew the path of the Gulf Stream, as a means of communicating the information graphically with nothing better at hand. In any case, by printing the current of the Gulf Stream about 1770 on a map drawn before 1720, Franklin, Folger, and the postal authorities violated the principles of good hydrographic maps that hydrographers were striving to uphold. No wonder, then, that captains scorned the map and its information and instead used De Brahm's *Atlantic Pilot*, which was constructed according to more modern principles and techniques. How could they trust the accuracy of the Franklin-Folger information if it appeared along with geographical data they knew to be in error? Indeed, if captains remained skeptical altogether about navigating in reference to the Gulf Stream it is not to be wondered at for on a 1783 version of the Franklin-Folger chart, the position of the Gulf Stream was rendered even more simplistically and imprecisely than on the 1768 original and contained several major errors in geographical matters, as in the location of Bermuda. Franklin's chart was far from being the "practical solution to a practical problem" that it has been labeled.[6]

Apparently Franklin did not understand what was wrong with his map. Like Halley before him, he deprecated seamen of technical ability who scorned the learning of

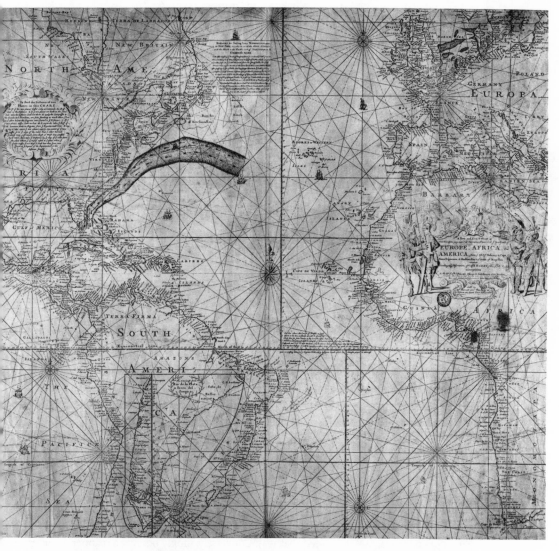

Figure 9 Franklin's chart of the Gulf Stream. On a map originally engraved by John Mount and Thomas Page, Benjamin Franklin superimposed the Gulf Stream, which he shows as a widening band in the West Atlantic. "A New and Exact Chart of Mr. E Wright's Projection, Europe, Africa and America from the Isles of Orkney to Cape Bona Esperance, Hudson's Bay to the Straits of Magellan." 96.0 cm × 84.5 cm. Phot. Bibl. Nat., Paris, Département des Cartes et Plans, Service Hydrographique de la Marine, portfolio 117, no. 7[1].

science and attributed to them backward habits of mind. He wrote, "Some sailors may think the writer has given himself unnecessary trouble in pretending to advise them; for they have a little repugnance to the advice of landmen, whom they esteem ignorant and incapable of giving any worth notice; though it is certain that most of their instruments were the invention of landmen."[7] Franklin nonetheless persisted in investigating the Gulf Stream, motivated as much by the belief that more knowledge about it might shake up the attitudes of mariners as by curiosity. In 1775 crossing west and in 1776 crossing east, Franklin used Fahrenheit's thermometer to determine whether the Gulf Stream was measurably warmer than the surrounding waters, in which case navigators could determine when they were traversing it if—and this was a very big assumption—the captain had a thermometer. This technique, moreover, would compensate for the inability of scientific investigators to locate a dynamic ocean current such as the Gulf Stream with precision on a map, a problem only recently solved with the aid of satellites.[8] Charles Blagden, physician to the army and a fellow of the Royal Society, also gathered information about the temperatures of Atlantic currents in 1776 and 1777, possibly without knowledge of Franklin's work, and published his findings in 1781.[9] Franklin gathered additional information in 1785 with the assistance of Jonathan Williams, his nephew, and the captain of the ship on which they traveled, Thomas Truxton; all three produced papers that included ideas on improving navigation generally and charts showing the Gulf Stream and various shipping tracks.[10]

These papers were published in the journals of scientific societies and circulated among the communities of scientifically minded men. There is little evidence, however, that any of the authors attempted to communicate his findings to mariners through their communication networks or tried to find out in what form mariners received information. The articles by Williams, Franklin, and others contributed more to early oceanography than to seamanship, yet they were presented as being of direct public benefit. Were these investigators so naive as to believe that navigators read the journals of scientific societies? Williams, for one, had no illusions on this score: "I think the observations of a mariner, are more likely to be attended to by mariners, than any instruction given by a landsman."[11]

Although the Franklin-Folger chart and its successors tell us very little about the Gulf Stream itself and are even unreliable sources about how much was known about the Gulf Stream in the eighteenth century, they are of interest for the light they shed on the development of cartography. Some maps with an avowedly utilitarian and practical appearance as graphic documents primarily communicated scientific knowledge. Simply because a mapper intends his map to have a practical application does not mean it will be used that way. Conversely, simply because a mapper includes information or techniques that eventually are included in everyday, utilitarian maps does not mean that practical concerns dominated his design. The story of these Gulf Stream charts reveals a latent and significant conflict between the uses of maps in science and in technology and confusion about the different standards that apply to each.

The scientifically curious and the practically minded often cast suspicion on each other's reasonableness, for reasons we can understand even if they did not. However hard it may be to separate theory from practice when many people are involved with both aspects of knowledge, this distinction should nonetheless be made, if only for its heuristic value. The practical man—the navigator or engineer—who cared little for theories about oceanography or geology nonetheless needed maps with consistent, reliable information about the seas and the mountains. The scientifically minded, who did not need to make a landfall, build a canal, or erect a fort, were more interested in using maps to illustrate hypotheses than in gathering the kind of data needed on utilitarian maps. Ironically, scientists, who emphasized accuracy in geodesy, were more casual when it came to landforms. Scientists used cartographic symbols of landforms as a kind of graphic shorthand to illustrate ideas they expressed in more rigorous terms in words and tables. Engineers and navigators evaluated maps of landforms more severely, as prudence dictated. Yet for that same reason they also had an incentive that scientists lacked to improve the thematic, analytical sophistication and graphic accuracy of maps of the seas and of mountains. Ultimately, maps of landforms improved not because of any advance in the natural sciences, but in response to utilitarian demands for more information recorded more accurately.

Contour Lines to Buache

Let us apply these insights to the introduction of contour lines in marine charts. Contour lines are isometric lines; these portray values that actually exist at points on the earth. Isometric lines can show surface temperatures, the declination of the compass needle, or elevation and depth. Isometric lines are to be differentiated from isopleths, which they resemble in appearance. Isopleths are "higher order, more complex, geographical concepts or abstractions that are a function of one element and space, such as density (e.g., persons per unit area), spacing (i.e., distance between), geographical ratios, correlations, and so on."[12] The values on which isopleth lines are based cannot actually exist at points on the earth. Isobaths are isometric lines that show depth below a water-level plane of reference. The advantage of isobaths is fundamentally a matter of legibility. When soundings are multiplied where they matter most, near the coastline, the numbers indicating the depth of water are crowded together on a chart. There is often a great difference in depth at two places that are very close together, but this difference, which may be of great consequence to the mariner, can be difficult to discern when indicated as numbers on a chart. (In any case, experienced mariners used several methods simultaneously, combining information from soundings, including both depth readings and the material of the seabed itself, with printed profiles of the coast, charts, and a richly detailed oral tradition that transmitted knowledge of landmarks.) Some hydrographers who simply printed sounding depths as numbers colored these figures differently according to the range of depth. Contour lines can make the irregular and invisible outline of the seabed appear more clearly on charts.

As far as can be determined, contour lines were first used in the eighteenth century by Count Luigi Marsigli, whose *Histoire physique de la mer*, published in 1725, truly opened the study of the sea to the world of science. Marsigli's book was only about the Mediterranean, not all the seas; nevertheless, it was still the first book devoted exclusively to marine science and the first to treat all the aspects of a single maritime region. It was based on two decades of research. Marsigli had been attracted to marine studies because he thought the Mediterranean Sea and the mountain chain of southern Europe were part of the same geological formation. He believed that, just as the seas are fed by rivers, the seabed is a continuation of mountain ranges. He further believed that the nature of the seabed could be inferred from the characteristics of coastal areas. Marsigli studied aspects of the sea that scientists knew little about: temperature, color, salinity, wave action, and currents. His book contained a map based on soundings that showed the outline of the continental shelf and a picture illustrating thirteen cross sections of the seabed. Marsigli used contour lines on his map of the continental shelf off the coast of France to distinguish between two levels of depth.[13] Marsigli talked with sailors about the sea but admitted that they were uncommunicative and appeared uninterested in his work. Perhaps Marsigli could not tell them anything they did not already know, although in unscientific terms; perhaps they simply resented him as an intruder in their world.

The use of contour lines for their practical value was the achievement of Nicolas Cruquius (1678–1754). A Dutch surveyor, Pieter Bruinss, utilized isobaths on a map of the river Spaarne in 1584, and Pierre Ancellin applied them to a map of part of Rotterdam and the Nieuwe Maas in 1697, but neither apparently had any followers or wrote about the technique in a discursive and formal way. Cruquius had made many soundings of the riverbed of the Merwede as an engineer responsible for dikes and waterways. To make the relief of the riverbed apparent, Cruquius connected points of the same depth with lines, at ten-foot depth intervals; he also showed the low-water mark, itself a contour. Given the shallowness of the water, he gathered enough data to add a dotted contour line at five-foot depth intervals. His map, designed in 1729, was printed in 1733.[14] In the Netherlands thereafter, however, this technique was used no more often in the next century than in other countries—which is to say scarcely at all. The Van Keulen family of hydrographers, for example, did not transfer Cruquius's technique to their own work. Each time the isobath was utilized for its practical value appears to have been an isolated instance, as if each cartographer had invented it himself. At about the same time (1729), a map was made of the Garonne River near Bordeaux, with numerous cross sections taken to show the depth of the river at different locations—another way of conveying information about a waterway that could be understood easily without instruction.[15]

The next contributor to the use of contour lines was Philippe Buache.[16] In 1734 Buache made a map with the profile of the ocean depths along a line between West Africa and Brazil and also drew an inset on a map of the ocean near the equator in which the Island Fernando de Noronha appeared, encircled by contour lines marking depths. In 1737 Buache presented to the Academy a map of the bed of the English Channel, based on soundings recorded on published charts (fig. 10). In his map of the Channel, Buache used contour lines most ambitiously. The points of soundings did not appear; rather, contour lines simply showed depth at intervals of ten fathoms. The map is also of interest for an inset, a profile or cross section of the Channel's bed along a line nearly equidistant from France and England that he clearly marked on the map itself (the cross section being divided along a vertical scale corresponding to the intervals between contour lines). Both the Atlantic and the Channel maps

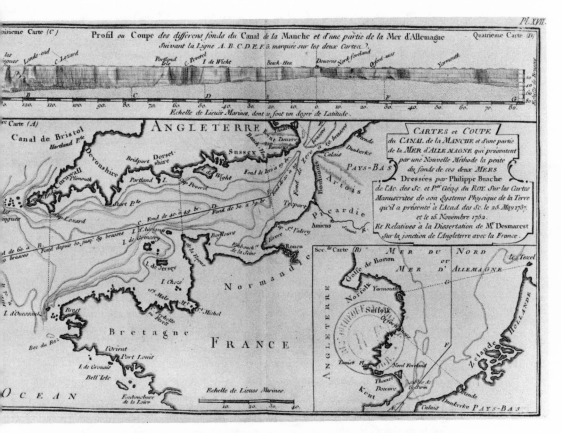

Figure 10 Map and cross section of the Channel. Engraved in 1752 from material drafted in 1737, this map by Buache shows contour lines for the Channel at intervals of ten "brasses." The line drawn in the middle of the Channel corresponds to the cross section at the top. "Cartes et coupe du canal de la Manche et d'une partie de la Mer d'Allemagne qui présentent par une nouvelle méthode la pente du fonds de ces deux mers." 40 marine leagues = 3.9 cm; 32.6 cm × 25.7 cm. Phot. Bibl. Nat., Paris, Ge. DD. 2091 (xvii).

were designed to illustrate Buache's thesis that the major continental mountain chains connect with each other underwater to compose a worldwide system.[17] Buache printed the Channel map of 1737 in 1752, together with a map of the world to illustrate "An Essay of Physical Geography, wherein it is proposed to present general views on what may be called the framework of the globe, composed of mountain systems that cross seas as well as continents; with some particular remarks on the different basins of the sea, and on its interior configuration."[18] The map of the Channel attempted to prove that the European continent and England are joined geologically.

Notwithstanding his use of contour lines, Buache had already grasped the limitations of two-dimensional flat maps to show relief. In concluding his essay, Buache proposed making a three-dimensional globe, one "that would allow the viewer, by lifting a layer representing the surface of the sea, to observe submarine topography, as well as the elevations above sea level on the continents."[19] A year later, at a meeting of the Academy on 22 December 1753, Buache spoke of that globe as being in the process of construction for display under the dome of the Luxembourg Palace. (This globe was never made, for lack of money.)[20] In the absence of detailed data on land elevation, Buache intended to deduce the heights of mountains from the gradients of the rivers that flow from them. And from his belief that the highest points on land were in the center of the continents, Buache inferred that the greatest depths on the floor of the oceans were farthest from land. On maps showing the rivers of Europe draining into the Atlantic or the Mediterranean, Buache had drawn two lines for each river basin, one representing the highest uplands determining the basin, the other the extent of subsurface waters belonging to it.[21] In 1757 Buache finished a simpler globe whose surface was depressed to show the depths of the seas, such that islands and submerged rocks appeared to be connected to mountain chains on land.[22]

These examples show that Buache used contour lines as only one of several techniques for conveying scientific information about the earth graphically. Buache may have been familiar with Cruquius's or Marsigli's work from the navy's hydrographic office. He may have invented the use of contour lines independently. Or he may have been inspired by the example of Edmond Halley's maps of magnetic variation, which used isogonic lines, and transformed that graphic device into contour lines to show depth and elevation. In that case Guillaume Delisle, Buache's father-in-law and a great map editor, would have provided the critical link between Halley and Buache, because Delisle had acquired much of Halley's work.

Buache wrote about Halley's maps in a letter to Jean Frédéric Maurepas after Delisle died.[23] From the letter we can infer that for Buache the use of contour lines and of isometric lines more generally in utilitarian, everyday charts and maps was premature. Edmond Halley had made maps of magnetic variation in the South Atlantic and of the world. These maps used isogonic lines, that is, lines connecting points of equal value limited to data on magnetic variation. Buache, in his letter to Maurepas, remarked that Halley's maps were not useful to mariners, as Halley had intended them to be, not only because Halley's theory of magnetic variation as a constant that did not vary from year to year in the same place was wrong, but also because Halley had tried to fit his observations into his hypothesis, thus committing many errors. Buache faulted Halley for failing to collect enough data. He observed that Delisle had gathered over 15,000 readings on magnetic variation from about 150 individuals.

Buache urged Maurepas, as minister of the marine, to have better isogonic charts made, with information collected routinely from navigators. He also composed a list of items for navigators to notice when recording compass headings, items that could be studied for their own value as well: whether magnetic variation appears to change irregularly or regularly; by how many degrees it declines; places where the compass points true north; estimates of longitude and latitude; direction and speed of current; time, strength, and direction of the tides; and speed and direction of the wind.

Buache, a half-century before Dalrymple and a century before Matthew Maury, had devised a scientific program for navigators to conduct in their own interest. Buache wanted to learn what relationship, if any, prevailed among these various aspects of the sea so that he would know what to emphasize in composing navigation charts.

This suggests that Buache perceived the applicability of isometric lines to marine charts early on but cautiously refrained from using them until he had acquired a coherent understanding of marine systems. He grasped that some scientific information on charts did not necessarily meet the needs of mariners. Buache realized that although advances in marine science and in seamanship were dependent upon and stimulated each other, the means appropriate to a diffusion of knowledge among scientists and seamen were different.

At this point Arthur H. Robinson and François de Dainville would shift from hydrographic to geological maps. Beginning with Buache's work of 1730–55, Robinson wrote, "We need trace no longer the development of the isobath as a navigational aid because it had become of age, so to speak. . . . Everything was ready, cartographically-speaking, for the isobath, as de Dainville so nicely puts it, to climb out of the sea, like Aphrodite, and become the dry land isohypse."[24] This view is inadequate on two counts. First, the use of isobaths on hydrographic charts was not yet the norm. And second, the use of contour lines to represent land elevation after 1750 owed little to the use of isobaths before then.

Routine Hydrographic Mapping

The isobath did not become common in hydrography for another half century and more. Only in the mid-nineteenth century were sailors able to make deep-sea soundings; Matthew Maury's chart of the North Atlantic (1853) showed over one hundred soundings greater than two thousand fathoms. Enough data on depths to make isobaths probably existed for many estuaries, coastal passages, and ports even in the eighteenth century, but in the absence of isobaths on their charts, sailors relied upon a traditional mix of techniques including soundings, coastal profiles, samples of the seabed, and verbal lore to guide their ships. Even without isobaths, however, French hydrography improved significantly in the eighteenth century.

The accomplishments of the French in hydrography stand in contrast with English practices. William Mountaine and James Dodson, for example, had the same idea as Buache, that isogonic charts should be made periodically to see how observations of magnetic variation in the same place vary over time, and in 1744 they began collecting observations from the logs of ships in the Royal Navy and the East India and Africa companies. However, many of these journals did not contain observations of variation; others contained observations but omitted longitude, making it impossible to determine where the readings were made. As a result, many observations had to be discarded or could be used only with great caution. Mountaine and Dodson published a "table of 50,000 figures which were adapted to every 5 degrees of latitude and longitude over the most frequented oceans," and a chart that never sold enough copies to cover production cost.[25] Needed but lacking in England was an agency that could standardize the methods of recording data gathered at sea and subsidize the cost of publishing up-to-date charts.

As a result, in eighteenth-century England the private mapmaking sector dominated hydrography as individuals with a flair for science tried to chart coastal waters. Hydrographers had to spend most of their time on land, not only because they lacked the support facilities to make extended cruises, but also and especially because they were handicapped by the absence of geodetic data about the coastline—information the French had gathered systematically by 1744. The importance of land surveys for hydrography is underscored by the fact that James Cook, the best British hydrographer before Robert Fitzroy and Francis Beaufort, learned surveying methods from two army engineers (Samuel Holland and Joseph Desbarres) in Canada.[26]

At least mariners had the good sense not to purchase charts in the private sector, which was not successful commercially in the field of hydrography. Lacking the capital to make good charts, map printers tended to reissue charts with few modifications for as long as the copper plates lasted, perhaps for a century (as we have already observed in the case of Franklin's first map of the Gulf Stream). Commercial success might not have enlarged the number of persons interested in making hydrographic charts in England, but failure certainly kept it small. The technical success of Murdoch Mackenzie senior in making a triangulation survey of the Orkneys between 1744 and 1749 inspired the authorities to back him in a grander scheme, a survey of the entire west coast of Scotland, which had never before been charted. Yet even this survey resembled other charts made with less sophisticated methods and was reprinted without correction for decades thereafter.[27] Private publishers, who plagiarized freely from each other, retained control of hydrography in the United Kingdom until the Admiralty established a hydrographic office in 1795—fifty years after Admiral George Anson and thirty years after Lord Howe had called for its creation.

As late as 1805, Alexander Dalrymple, the first hydrographer, was still able to state that "it is a great discredit of this country that few places are determined in latitude precisely and still fewer in longitude; notwithstanding the perfection to which Instruments and Chronometers are now brought."[28] Once again we are reminded that the introduction of new techniques and their impact on professional practice were far from synchronous. Blinded by the achievements of John Harrison and Larcum Kendall in horology, we make the simple mistake of assuming that because the longitude problem had been solved in England in the middle of the eighteenth century, all was put right. Along with chronometers, Dr. Gowin Knight's azimuth compass, John Bird's astronomical quadrant, Murdoch Mackenzie's station pointer, and John Hadley's octant permitted a navigator to measure astronomical angles and the angles between his vessel and objects on land reliably.[29] Given the advances in instrumentation achieved in England, it is astonishing how slowly better instruments were incorporated into hydrography. A score of years lapsed between the time when most of these instruments were perfected and the publication of Dalrymple's and Mackenzie's treatises on marine surveying, the works that presented routine procedures for their use.[30] Chronometers were not standard in the Royal Navy until the 1840s.

In England, James Cook at sea and William Roy on land developed excellent methods and high standards before the Hydrographic Department and the Ordnance Survey were established—but to diffuse their methods and sustain their standards, these institutions were absolutely vital. Some might say that in France, because naval traditions were shallower than in England, public and institutional initiatives were more necessary, but this misses the point—that whatever the degree of interest in naval affairs in any country, only public, institutional initiatives sufficed to promote and sustain a scientific approach to navigation and to hydrographic charting.

The differences between England and France in the development of hydrography were a function of institutions and attitudes far more than of techniques and inventions. The French established a single hydrographic agency, the Dépôt des Cartes et Plans de la Marine, in 1720; the English established a comparable agency in 1795. And the French hydrographic office was only one of several institutions devoted to maritime science in eighteenth-century France. There were schools of hydrography in Brest, Toulon, and Rochefort, and the Compagnie des Indes maintained its own school in Lorient. These schools, however, acted as a conservative influence, not a progressive one; traditional methods of navigation were still preferred to reliance upon scientific instruments, partly because these were still few in number and on trial, and partly because the instructors preferred familiar methods.[31] As so often happened in France, a parallel set of institutions emerged that were more eager to adopt new techniques. The *Traité de navigation* by Pierre Bouguer, published in 1750, was the first modern text on the subject and the first written by a scientist. It made students of the subject less dependent upon schools and apprenticeships. Shipbuilding and naval architecture were revised by Henri-Louis Duhamel-du Monceau, who even assembled a research group to evaluate various industrial techniques in shipbuilding.[32]

In 1752 the Académie de Marine was established in Brest; in 1769 it received royal letters-patent.[33] The members, including Bouguer, J. B. d'Après de Manevillette, Pingré, Lalande, Chabert, and Borda, examined topics in naval science, analyzed logs, evaluated proposals and inventions, and conceived of a "dictionnaire de marine," part of which was published as the "marine" section in Panckoucke's *Encyclopédie méthodique*. This academy prepared the scientific programs for several voyages of discovery; some of its members sailed with Louis-Antoine de Bougainville and the Comte de La Pérouse. It also supervised the testing of chronometers by Pierre Le Roy and Ferdinand Berthoud, chronometers that came much closer than any of Harrison's early pieces to approximating the modern device.[34]

Symbolic of the achievements of the French in naval science were the voyages to observe the eclipses of Venus in 1761 and 1769. Halley had predicted these events and explained their importance to the measurement of the distance between the earth and the sun. For the transit of Mercury in 1753, a map had been made showing those parts of the earth from which the transit could be seen. Late in 1752 or early in 1753, Joseph-Nicolas Delisle made a map and wrote a report on the transit of Venus for 1761 with similar information. Each hemisphere was divided into three regions, indicated by shading: one in blue represented that part of the earth illuminated by the sun at the moment of Venus's entry upon its disk, another in red represented that part darkened by the full time of Venus's transit across the sun, and the third in yellow represented that part from which the moment of Venus's exit could be observed. The first and third regions were further divided to show where entry or exit would occur at sunrise or sunset. In this manner every zone and degree of visibility of the transit for the entire globe could be distinguished. Joseph Lalande composed a map showing the effect of parallax on the times of entry and exit at different places on the earth. Delisle's original was submitted to the Academy on 30 April 1760 and figured in the attempt of the French to organize voyages to the South Seas, where conditions for observation would be optimal.[35] In 1761 Guillaume Le Gentil went to Pondicherry, Pingré to the Mascarenes, and Jean Chappe d'Auteroche to Siberia. The English were too preoccupied with the Seven Years War to send observers to Saint Helena and to Hudson Bay, as previously arranged. For the transit of 1769, Chappe d'Auteroche went to California and Bougainville to the Moluccas; Le Gentil was still in Pondicherry, where he had remained after the British captured the colony in 1761. This time the English made a showing in the South Seas, sending Cook (whose charts were the first to run longitude from the Greenwich meridian). These voyages benefited from improvements in navigational skills, in naval medicine (in 1740–41 Anson lost half his men to illness, but in three years Bougainville lost only seven, and in two years La Pérouse lost none), and in naval architecture (enabling larger ships of shallower draft to approach land). This was the last chance for navigators to make major discoveries and the first chance for them to explore since the writings of Buffon and Rousseau. Joseph Banks traveled with Cook, and Matthew Flinders carried the *Encyclopaedia Britannica*. These voyages of 1761 and 1769 brought scientists and mariners together as partners in a common venture. On his later voyages, La Pérouse traveled with astronomers, naturalists, and a translator. He also carried Berthoud's chronometers, over a thousand books, and a large stock of maps. The scientific program for his voyages, written by Charles Pierre Claret de Fleurieu, ran to nearly one-hundred pages. Occasionally there was friction between officers and scientists and also among the scientists themselves over who was in charge, and thus over the priority of investigations to be undertaken. But the scientific purpose seems to have overcome narrow-mindedness. Joseph Banks let La Pérouse travel with Cook's compass on what proved to be his last voyage, and when he vanished what was probably the world's first international rescue mission was launched.

Behind these spectacular voyages, and less visible than the heroic captains who led them, were the institutions devoted to the promotion of naval science and their dedicated, professional staffs. Representative of France's leadership in hydrography was Charles Pierre Claret de Fleurieu (1738–1810). A junior deck officer in the Seven Years War and a talented watchmaker, he commanded the *Isis* for the tests of

Berthoud's chronometer and wrote an account of that voyage. He then gave up a life at sea for a desk job as deputy inspector of the Dépôt des Cartes et Plans. One of his major tasks was to make charts as accurate as chronometers—that is, to within half a degree of longitude. This involved corrections of sixty, eighty, and even one hundred leagues on many charts, and hence the evaluation of nearly everything that had been printed or drawn without the benefit of modern scientific methods. After the Compagnie des Indes expired in 1769, Claret de Fleurieu made the Dépôt the principal repository of hydrographic charts in Europe. In 1790–91 he became minister of the marine, probably the first time someone with his professional training had risen to such a position.[36]

Claret de Fleurieu's work in hydrography was surpassed by that of Charles François Beautemps-Beaupré, whose coastal surveys in the early nineteenth century, made with Borda's reflecting circle and advanced sounding devices, are among the most elegant and sophisticated ever made. His chart of the North Sea included a modified use of isobaths. Basing sea level upon observations of the ebb tide at equinox, Beautemps-Beaupré determined depths. He distributed soundings into three categories (depths to ten feet, from ten to sixteen feet, and below seventeen feet). On the map he separated them with dotted lines and shaded each area in a different tone. On his chart of the coast from Calais to Ostende, he indicated those portions of the land that were visible from the sea and displayed prominent shore objects in the aspect that a mariner would perceive them as an aid to navigation—techniques by then two centuries old. What was new about Beautemps-Beaupré's use of these techniques was that he rigorously excluded from his charts elements extraneous to the seaman, such as objects on land near the coast that would not be visible from a ship.

The Dépôt was effective because the people involved in its work were determined to make it so. This required two kinds of investment, one intellectual, the other financial. In one of his first speeches to the Academy after his entry in 1729, Buache described his and the Dépôt's methods. There were some four thousand charts and maps, six hundred to seven hundred memoirs and letters, and four hundred logs in the collection at that time. Each observation relative to a track, a landfall, or a distance between two points was checked against known astronomical data and valid charts. Each observation was then transcribed onto a slip of paper called a bulletin, on which its source was also recorded; all the bulletins were then classified by subjects, called divisions. Various bulletins were then transposed onto maps, with colors and symbols to distinguish the different sources. These methods did not vary much over the life of the Dépôt in the eighteenth century—they were very good right from the start.[37] Let us see how the men of the Dépôt worked. Jean-Nicolas Buache de la Neuville, Philippe Buache's nephew, came upon a report of the transit of Venus of 1769 written by four Americans and published in the first volume of the *Transactions of the American Philosophical Society*. Buache realized that the place where they stationed themselves, Cape Hinlopen in Delaware, was mislabeled as Cape James on the chart of the Delaware River published in 1777 in the second edition of the *North American*

Pilot. By comparing that map with this report, he further realized that on the map, a lighthouse situated to the south of Cape Hinlopen (and mislabeled Cape James) in fact was to the north. He then recommended changes to the copperplates for charts of that region.[38] All in a day's work, perhaps, but only if one has many days to spend at such tasks.

In the 1670s and 1690s, the Cassini-Picard team had corrected the outline of France's coasts, thereby dramatically demonstrating the superiority of geodetic fieldwork. In much the same way, the hydrographers of the Dépôt used the observations of navigators and astronomers to correct the outlines and dimensions of the Mediterranean Sea and the Gulf of Mexico in the 1730s and 1740s.[39] They used the Paris meridian to facilitate comparisons between observations by navigators at sea and by astronomers in Paris against a common time standard. Only by gathering data from reliable observers and by taking voyages themselves could hydrographers continue to revise and perfect their charts. Buache and Dalrymple designed tables for mariners to fill in with observations that would standardize and simplify their work, and they exchanged information so that hydrographers in both countries might follow a common format. Dalrymple recommended only minor changes in the logbook, however, "since men are guided so much by habit in marking it, that the errors committed would be infinite, if their habitual form was changed." He proposed adding an entry for winds, and he developed his numerical wind scale from 1 (faint air) to 12 (storm) for such tables. The oldest known version of this wind scale is the one he sent to the Dépôt in 1779.[40] However, Dalrymple's scale received so little notice anywhere that until his memorandum was discovered in the Archives Nationales there was no proof that he had in fact developed a wind scale, and Francis Beaufort was credited with the idea.

Not every cartographer was qualified to work in the Dépôt. In 1772 Lalande received a request from J. A. Rizzi-Zannoni to support his candidacy for a vacant position there. Lalande preferred Rigobert Bonne, citing his ability to compare and analyze logs and make good maps, as well as his reliability and integrity. Lalande examined a map of the Channel by Rizzi-Zannoni: his mistakes included misplacing geographical details, failing to draw conclusions from the available evidence, and making maps of a scale inappropriate for navigation. Lalande showed good judgment when he stated that Rizzi-Zannoni lacked *bonne foi* (good faith).[41] Clearly, efforts had to be made—and were successful—to keep incompetent and dishonest people out of the Dépôt. A post there was not a sinecure; professional standards counted more than patronage.

In the operation of the Dépôt, financial matters were as important as intellectual standards. This was true not only because mapmaking of this kind was costly, but also because, to encourage navigators to buy and use better charts, the state subsidized their purchase price and did not try to recover its initial expenses through sales. Buache arranged in 1737 for the Treasury to reimburse any loss incurred by selling maps to distributors below production costs. The price to the retailer was to be low enough so that he could set the final price to the client at an amount a third less that of older maps yet still produce a profit.[42] Costs of producing an engraved copper plate fluctuated between five hundred and one thousand livres, depending on the size of the plate and the amount of detail; paper, printing costs, and coloring for one hundred sheets could run another thirty to forty livres. (Buache estimated that 10 percent of the printed sheets would be defective.) To be competitive, the Dépôt's maps had to sell at retail for twenty sols if on paper or for between fifty sols and three livres if on parchment, which held up better at sea. Because so many hydrographic maps had to be updated so often, production costs were high.

The Dépôt was successful because it was conservative in the sense that its maps contained verified data and excluded conjectural or unverified information; but its success also made it conservative, in the sense that people came to expect a certain continuity of method and style. It is worth remarking that Buache's most innovative work in hydrography came predominantly during the first half of his career (before ca. 1755). When the Revolution came, Claret de Fleurieu considered the implications of the metric system for hydrography and concluded that change could come only slowly, after a long period of transition. He knew that so long as older books remained in circulation, and so long as the rest of the world continued to use traditional measurement units, charts would have to conform to tradition. People trained in the metric system would still have to use maps and read books that did not take the metric system into account. Claret de Fleurieu thought that hydrography needed a standardized nomenclature system more than a standardized measurement system.[43] Quite simply, hydrography could not evolve quickly.

Du Carla, Dupain-Triel, and Contour Lines

There was a gap after Buache in the transmission of knowledge about contour lines. When the history of hydrography in the eighteenth century is considered in all its dimensions, the introduction of contour lines recedes in importance. Their use in midcentury was still inopportune, and their absence did not retard the introduction of more scientific approaches to navigation and hydrography. It is incorrect to suggest that there was a direct path from Buache to those who used contour lines after him to represent depths in the water and heights on land. Contour lines were reinvented by someone else after Buache, as if his earlier work had never existed. The systematic use of contour lines owed more to Marcellin Du Carla, a Languedocian geographer, than to Buache.

In 1782 a publisher and mapmaker named Jean-Louis Dupain-Triel, Sr., paid for the typesetting and engraving of a manuscript by Du Carla, entitled *Expression des nivellements, ou Méthode nouvelle pour marquer rigoureusement sur les cartes terrestres et marines les hauteurs et les configurations des terreins*. Six hundred copies were printed; Du Carla received the first 150 and Dupain-Triel took the rest, presumably to sell. This was the first fully developed exposition of contour lines. Du Carla began by commenting upon the incompleteness of maps in respect to knowledge of sea depths and mountain heights. His goal was to achieve a true picture of the earth's configuration. Man cannot see the world as it is by looking at it, he wrote; reality emerges as a synthesis of information on paper. That man could now do this was, in Du Carla's opinion, a sign of the superiority of eighteenth-century culture over earlier times.

Du Carla argued that contour lines made it possible to add information about height and depth without overwhelming a map with information. He did not believe that isobaths and isohypses were exactly the same, because land forms and sea forms are themselves dissimilar. Moreover, he pointed out that we cannot ever see the seabed for ourselves, whereas we can check the representation of a mountain on a map against the mountain itself. Nevertheless, for practical purposes, he discussed the two kinds of lines together. Du Carla used sea level as point zero (at a time when military engineers certainly did not), but he understood that even sea level is arbitrary, since it varies according to seasons, tides, and latitude. His idea—to determine the level of the sea where rivers enter it and then average these observations— was rather in advance of scientific and technical work for his day. He also wanted to use sea level because he considered mariners the group that would benefit most by the use of contour lines. Du Carla suggested that every tenth contour line on a map (representing ten toises if the lines are at one-toise intervals) be thicker, to make it more visible. He consciously understood this as an instance of decimalization but allowed for the use of a duodecimal division if more convenient.

Du Carla expressed great conviction about the superiority of maps based on measurement of height and depth. With contour lines on a map, he wrote, it becomes possible to visualize what can and cannot be seen from any one point. Contour lines are a language, like Greek or algebra or geographical symbols. From them a three-dimensional image of space can be constructed in the mind, far more easily than if many numbers were scattered across the surface of the map. With contour lines it becomes possible to measure height and depth on a map. He was a meliorist, convinced that, over time, accurate measurements of height would become as common on maps as accurate geodetic calculations. Du Carla hoped that landowners would make contour maps of their estates and that scientists and engineers would eagerly gather and share information about elevation, so that a national map could be constructed. To encourage others, Du Carla announced that he was prepared to publish a two-volume treatise on the theory and method of measuring elevation for maps, to be illustrated with three large plates: one illustrating all possible examples, a second showing France with contour lines, and a third showing the depths of the English Channel in greatest detail. It must be presumed that an insufficient number of subscriptions (at thirty-six livres each) were made, for du Carla soon retired from active geographical work.

That he ever came to write the one book that constitutes the sole basis for his reputation is itself remarkable. Du Carla claimed he had worked out the idea of contour lines as early as 1765, and he began expounding it to other people. In 1771 he consulted with Henri-Louis Duhamel du Monceau, Jean-Paul Grandjean de Fouchy, and Buache himself in Paris. He also presented his work orally to the Academy on 4 May 1771. Grandjean de Fouchy asked Du Carla to name the commissioners who were to evaluate his presentation; Du Carla said he would accept the verdict of the

Academy as a whole. In the end, Charles Messier was chosen to head the panel, which included Buache. Their report, dated 11 May 1771, was unenthusiastic; they believed contour lines were too difficult for the average map reader to understand. Du Carla recorded that Buache never contested his claims to have been the first to explain the theory and practice of contour lines. Du Carla returned to Languedoc soon thereafter, and he dropped the matter until 1780, when on a visit to Geneva he met with requests for explanations. Back in Paris in 1781, Du Carla had one map made with contour lines and experimented with hydrographic charts. Obviously it was only on his visit to Paris in 1781 that Du Carla took the formal step of securing the publication of his work.

Du Carla, a man of limited education compared with most of the scientists close to the center of geographic and hydrographic activity in Paris, understood the geometric properties of contour lines as conic sections and understood their geographic characteristics as true isarithms. Du Carla's understanding of contour lines was a function neither of his study of a specific scientific problem nor of his mapping activity, but was rather an idea that he grasped abstractly and then tried to work out in its practical application. This rather extraordinary performance highlights not only the occasionally improbable circumstances that attend upon important developments, but also how much less is sometimes achieved by people working in far more favorable conditions. If Du Carla's own version of his story is correct, then it substantiates the gap I believe existed in the transmission of contour lines as a cartographic technique. Buache had never taken the step Du Carla took, to explain and encourage the use of contour lines *independent of any particular theory about oceans and mountains.*

Yet the transmission was not secure even with the publication of Du Carla's book. In 1791 Jean-Louis Dupain-Triel, Jr., came out with the first map of France based on Du Carla's methods—the map Du Carla intended to make himself—based on readings of elevations from naturalists' publications, the Cassinis, and engineers of the Ponts et Chaussées (fig. 11).[44] It included a cross section taken on a line between the Rhône below Vienne and the Gironde. In the Year VII (1799) Dupain-Triel republished his contour map of 1791, changed its title, and substituted a table of some forty places in France in order of their altitude for the cross-section profile.[45] Each time, Dupain-Triel hoped his map would inspire the revolutionary government to establish measurement teams in each department. In 1802–3 a government commission on mapping techniques and symbols authorized limited use of contour lines only. By 1804 Dupain-Triel's optimism had given way to disappointment, as he noted that his earlier work—and by implication Du Carla's as well—had been ignored.[46] He repeated the case for making better maps with contour lines, whether on a world map for the study of mountain chains that girdle the earth or on a local map for the design of a road or bridge. Dupain-Triel restated Du Carla's idea when he wrote that contour lines are the language of elevation, but a language with only one letter. Reading the lines is every bit as concrete as looking at a landscape illustration, but, he wrote, it is better to study the map than to look at nature, not only because the appearance of nature

Figure 11 Contour map of land elevation. Drawn and engraved by Jean-Louis Dupain-Triel, Jr., in 1791, this was the first map of its kind for France. The inserts in the lower left show a cross section from the Gironde to the Rhône. Note how much more detailed the river valleys are than the mountains from the Juras to the Mediterranean. One large Alpine mass including Mont Blanc appears distinctly to the south of Lake Geneva, but it is not convincingly accurate. "La France considérée dans les différentes hauteurs de ses plaines: Ouvrage spécialement destiné à l'instruction de la jeunesse." 1:2,164,000; 53.5 cm × 46.0 cm. Phot. Bibl. Nat., Paris, Ge. D. 15126.

tends to confuse, but also and especially because the principles of earth science can be synthesized only out of the study of earth forms on maps.

Dupain-Triel and Du Carla were both interested in contour lines because they viewed terrestrial and terraqueous phenomena as complementary aspects of one physical earth. The same line could be used to indicate features below and above the sea level, on land or underwater. They were above all interested in the relation between mountain and river, especially because the study of their relation could be applied to the design of major public works such as bridges, roads, and canals. This was also a concern of some other scientists and engineers in the late eighteenth century whose maps and writings will be considered in chapter 5.

Contour lines were first applied to mountains without regard to the precedents of their use in hydrography or to the ideas and models set forth by Du Carla and

Dupain-Triel. What cartographic conventions were used for mountains before contour lines, how did contour lines emerge in the context of mapping mountains, and what relation existed between scientifically minded and practically minded mappers of mountains?

Four

Maps of Mountains

Theory and Practice

Scientists and engineers concentrated upon mapping either landforms or the seas; only a few—Buache, Du Carla, and Jean-Baptiste Meusnier most notably—took an interest in both. The process of mapping mountains resembled hydrographic mapping in one critical respect: persons whose interest was largely scientific approached their subject differently than persons whose interest was predominantly practical. The mapping of mountains, no less than the mapping of seas, illuminates some of the differences between science and engineering.

Several factors limited the use of graphic forms such as maps, cross sections, and topographic drawings or engravings to convey information about mountains.[1] The length and verbosity of so many engineers' reports and naturalists' accounts implies that a high value was placed on literary expression. The attraction of verbal over visual media may have had much to do with a preference for classification and enumeration; it may also have reflected the uneven drawing ability among professionals and fieldworkers, who could use words more precisely than graphic forms. Not surprisingly, early modern maps of mountains conformed to a plainsman's view of them, which emphasized their mass, height, and irregular shape as perceived from a distance and from the ground looking up. That mountains were usually drawn in pictorial form, in oblique perspective, and with very similar details made sense when most people never got close enough to distinguish their individual characteristics or to see how their aspect changed with one's position. To be sure, there was the conceptual problem of collapsing the three-dimensional structure of the earth's surface onto a two-dimensional surface, but as an obstacle to verisimilitude this would have affected written and graphic forms equally. Nevertheless, many traditional graphic design practices were retained long after more effective ones had become available. Scientists were at a greater disadvantage than engineers. Scientists, who needed the services of a professional cartographer, engraver, or illustrator to make printed images for their publications, may have been frustrated by problems of communicating ideas and supervising design and production. The cost of making and printing illustrations of any kind must have been a factor in determining the kind and function of maps in scientific publications. Engineers were more likely to make maps themselves to accompany unpublished reports.

Although scientists did less than engineers to reform and perfect the representation of mountains on maps, maps were important to the formulation of theory in the earth sciences. The relation between such features as river basins, oceans, mountains, and plains and the processes by which such features had taken shape over time were topics appropriate to mapped images. At times maps even seemed to suggest a theoretical explanation of phenomena. As evidence accumulated and as theories became more complex, maps became more thematic and detailed—though whether this parallel evolution was coincidental or speaks of a deeper relationship is unknown.[2]

Yet the application of contour lines to mountains did not come about in the context of natural science. This achievement was the work of French engineers. Engineers and scientists emphasized different things about mountains in their maps and developed different techniques to display them. For the most part, engineers and scientists did not exchange ideas or techniques. (To the extent that many of the best maps made by engineers were used by the military, an exchange would have been difficult.)

Engineers and scientists traveled in mountainous regions for different reasons.[3] By and large, the engineers traveled on orders, to map areas through which armies might travel or in which fortifications might be constructed. Scientists tended to work by themselves (although accompanied by guides and porters) to satisfy their curiosity. Neither group had much to do with the populations of the districts where they traveled, a pattern somewhat reminiscent of Marsigli's difficult relations with Mediterranean mariners. The role engineers played in perfecting techniques to represent mountains on maps was more important than that of scientists, whose approach lacked rigor and appeared amateurish by comparison. To scientists, maps were of secondary importance and functioned chiefly as surfaces on which facts were recorded—not as images to be analyzed and interpreted, but as composite devices to present analyses and interpretations in graphic form. To engineers, maps were documents that provided the primary basis for interpretation and analysis. Engineers cared more than scientists about perfecting the accuracy of maps of landforms because, quite literally, the success of their projects depended on the reliability, legibility, and accuracy of maps.

Because engineers and scientists mapped mountains for different reasons and with different techniques, their work will be considered separately. Scientists tended toward geological mapping, which involved the location of mineral deposits, sub-surface strata, and theories about the origins and evolution of landforms. Engineers were more interested in the accurate depiction of landforms as these related to fortifications and military planning and to the construction of roads and canals.

The importance of theory to an understanding of mountains had been proclaimed early in the eighteenth century by Louis Bourguet (1648–1742). Bourguet, son of a Protestant merchant in Nîmes who moved to Switzerland upon the revocation of the Edict of Nantes, taught mathematics and philosophy in Neuchâtel. In a book published in 1729, he outlined a program of study about the earth's land surfaces for scientists to undertake.[4] Bourguet focused upon the general pattern of mountain chains on the earth's surface. Scientists had not made much progress in earth science, he wrote, because they did not pay enough attention to the interrelations between the most important phenomena. Bourguet believed that progress would come as geometry and a knowledge of structure were applied to the study of geography. Obviously, no one person could provide the necessary synthesis of knowledge. But taking the Cassini geodetic survey as a model, Bourguet hoped that many observers, under the protection of many princes, would be able to travel all over the world gathering the data for a theory of the earth. But teamwork, cross-fertilization among scientific disciplines, and the accumulation of data by careful observers with reliable instruments were not present in earth science until the early nineteenth century. Until then, funding for major expeditions and publishing ventures was irregular and inadequate when it was available at all. Advances in knowledge were achieved by self-motivated individuals working independent of each other, more often than not with an eye to discrediting someone else's theory. As Alexandre-Charles Besson wrote a generation after Bourguet, "On a recours à des imaginations, à des systèmes . . . au lieu qu'il faudrait voir, examiner, monter, descendre et beaucoup fatiguer."[5]

As a result, in the context of natural science, cartographic techniques to represent mountains evolved as a function of theoretical differences. In 1746 Jean-Etienne Guettard (1715–86), a naturalist by avocation, presented the Academy with a "memoir and mineralogical map of the nature and distribution of the lands that traverse France and England."[6] The map accompanying the memoir, made by Buache, displayed the location of mineral deposits in France by symbols placed on the map in the general area where these deposits were found (spot distribution). The spot distribution method satisfied Guettard, but it did not "suggest how a localized clustering of a particular symbol might reflect a (literally) underlying uniformity in the bedrock. The symbols denoted scattered *points* about which information was offered; they did not connect those points into any kind of pattern."[7] Guettard was interested in showing the division of the earth into broadly differentiated zones, based upon comparisons of the "different quantities of key substances found in each region he traversed or read about."[8] The continuity of these *bandes* or zones across the Channel was Guettard's contribution to the theory already put forward by Buache about the submarine connections between the continents. But the spot distribution technique did not illustrate this clearly. Guettard intended this map as a preliminary survey that would present information and demonstrate a cartographic method at the same time.

On maps illustrating his own theories, Buache used a more complex and appropriate technique. Buache made several maps of the earth showing mountain ranges on land and under water as continuous, shaded, caterpillarlike lines. This technique emphasized the continuity and homogeneity of mountain chains at the expense of individual mountains. Buache placed mountain ranges on his maps where he believed they ought to exist if his theories were valid.[9] He also interpreted watersheds as a function of mountain ranges, and he intended his maps of the world to be colored to show which portion of a continent drained into the Atlantic, Pacific, or Indian ocean. He made maps of France and of Languedoc to illustrate his theories in greater detail. On the Languedoc map, he indicated the range dividing the Loire, Garonne, and Rhône valleys from each other by highlighting it with red dots (figs. 12, 13).[10] Buache's approach to geography suggested an application to the selection of sites for roads and canals, but his cartographic technique was not detailed enough for engineers to use with confidence.[11]

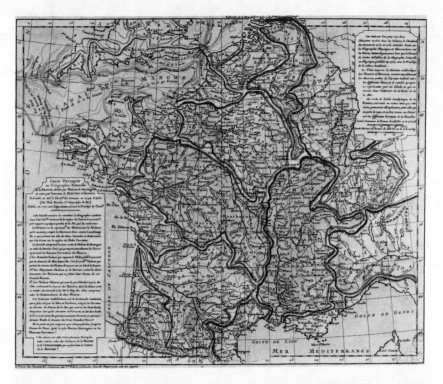

Figure 12 France divided into river basins. Drawn by Philippe Buache in the 1740s and 1750s and printed in 1770, this map shows parts of England, the Low Countries, Germany, Switzerland, and Italy and all of France. Note the inclusion of contour lines for the Channel but their omission from the Atlantic and the Mediterranean. Buache wanted to show transcontinental mountain ranges as the primary structure of river basins. These ranges appear as thickly shaded lines. "Carte physique ou géographique de la France, divisée par chaines de montagnes." 1:2,777,775; 45 cm × 57 cm. Phot. Bibl. Nat., Paris, Ge. DD. 5400, no. 4.

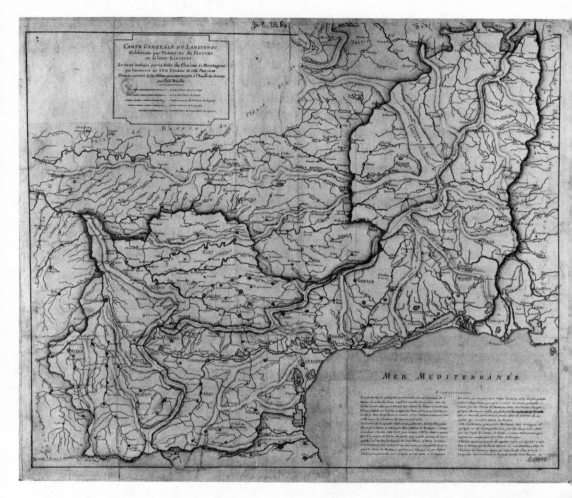

Figure 13 Languedoc divided into river basins. Drawn by Philippe Buache in 1766 but never printed, this map shows the same kind of information as Figure 12 but in greater detail for Languedoc. This map also shows the extent to which towns were situated on rivers and streams, the essential preindustrial source of power. "Carte générale de Languedoc. . . ." 1:435,730; 91.5 cm × 74.0 cm. Phot. Bibl. Nat., Paris, Ge. B. 2384.

When the Cassini survey maps became available, Guettard was able to free himself from dependence upon Buache. Guettard collected evidence by direct observation, from published sources, and from correspondence, which he depicted on Cassini sheets. In 1763 he began to work in the field and in the laboratory with Antoine Lavoisier (1743–94). Lavoisier also recorded much information onto Cassini survey sheets. Both men were aware of various geological theories, but Guettard's methods and concerns did not demand such precise measurement as Lavoisier's. Lavoisier's stratigraphic ideas were oriented toward the central debates in geology at that time, whether the action of water or fire effected the changes that produced the earth's surface, and whether this action was continuous and slow or violent and sudden.

In 1766, Lavoisier and Guettard secured from Bertin, then director of mines, approval and some funds to begin publication of an *Atlas minéralogique de France*; Jean-Louis Dupain-Triel, Sr., later Du Carla's publisher, began engraving the Vexin and Valois sheets. Lavoisier's and Guettard's different scientific interests affected the layout of each sheet. On the left was the legend denoting Guettard's spot distribution symbols; on the right were Lavoisier's idealized cross sections, which were the first pieces in what he hoped would be a study of France's crust during several epochs. In between was a map based upon the Cassini survey on which the spot distribution symbols appeared. Lavoisier might have eventually developed a means for combining subsurface cross sections with surface patterns and for relating these to elevation data gathered assiduously with a barometer (which supposedly indicated information about the limits of the sea at some remote time). He had the idea of linking all spot distribution symbols to show the size and extent of each deposit and their points of intersection.[12] But funds were reduced after 1772, before the team of Guettard and Lavoisier had been able to blend their techniques and interests. In 1777 Antoine-Grimoald Monnet, a civil servant in the Department of Mines interested in locating mineral veins, took charge. Monnet thought that Guettard's spot distribution technique alone sufficed; he eliminated Lavoisier's type sections. Afraid his own position would be cut in an economy move, Monnet hastened to bring out as many sheets as possible. By 1780, 30 maps out of a projected 214 were published, based almost entirely on a reduction and simplification of earlier fieldwork accomplished by Lavoisier and Guettard separately and together. Whatever its limitations, the *Atlas* was nonetheless remarkable for its consistency. That even 30 sheets appeared was a landmark event in geological mapping.

Lavoisier's immediate reaction to cuts in funds had been to suggest a procedure deviating so far from the standards of accuracy he normally supported that it shows how badly he wanted to continue the project. Lavoisier tried to persuade Henri Bertin to reduce the scale of the maps and to compile their geological content on the basis of questionnaires completed by curates, naturalists, and the gentry. The errors that would inevitably appear, Lavoisier thought, would be corrected in future editions. This proposal expressed a naive view that such a complex project could be maintained over many years, perhaps decades. Bertin rejected Lavoisier's proposal.

The work of Guettard, Lavoisier, and Monnet up to 1780 demonstrates first the importance of good geodetic surveys to the development of more specialized forms of cartography—as important in geological as in hydrographic mapping—and second the extent to which geology and geological mapping were, as Rapoport says, in "enough of a state of confusion and change to permit and indeed to invite the individual geologist to treat the science, within broad limits, much as he saw fit."[13]

With the work of Nicolas Demarest (1725–1815), the fate of an atlas of geological maps in France became more complicated.[14] In 1751 Demarest wrote an essay on the geological connection between France and England that received first prize from the Academy of Amiens. To illustrate Demarest's thesis, Buache revised and enlarged his earlier maps of the English Channel on which contour lines appeared. Demarest was primarily interested in his essay to discuss the action of the seas as the slow and regular process by which the land isthmus between England and France was destroyed. Buache's maps served Demarest's purposes well. The smooth and deep profile of the Channel, highlighted by contour lines, excluded the possibility that the isthmus had been destroyed suddenly and violently, as by volcanic action. (It is an interesting comment on the traveling habits of scientists that Demarest did not see the ocean until 1761.)[15] After that, Buache and Demarest went their separate ways, as did Buache and Guettard. Demarest became interested in Guettard's suggestion of 1751 (verified by Trudaine in 1763) that the *puys* of the Auvergne are extinct volcanoes, a point of direct interest to anyone interested in the aqueous/igneous dispute about the origins of rocks. Demarest continued making observations in the Auvergne with François Pasumot, a geographer who also collaborated with Guettard in the 1760s. Demarest worked with Pasumot in 1764; Pasumot was alone in the Auvergne in 1765, while Demarest visited Italy; and the two men joined together again in 1766.[16] They produced a geological map of the southern, Mont-Dore section of the Auvergne that was published in 1779; Demarest's complete map of the Auvergne was published posthumously in 1823. On his map of 1779, Rudwick states, Demarest "used a re-markably subtle range of engraving techniques to depict volcanic cones, lava flows, and isolated outlines of older basalts . . . as a means of displaying the evidence for a temporal reconstruction of the successive 'epoques' of volcanic activity that the rocks recorded."[17]

Demarest's texts about geological matters during the 1780s have received far more attention than his conflict with Lavoisier over a proposal to revive and continue the work begun with the *Atlas*. Demarest and Lavoisier were both civil servants and were brought together into a consultative council on agriculture perhaps even more as bureaucrats than as scientists. In 1785 a new Committee on Agriculture was created within the Contrôle-Général des Finances to respond to requests for information and advice from farmers and to evaluate proposals and inventions.[18] At its first meeting Nicolas Demarest, then an inspector general of manufactures, presented a proposal for an inventory of France's geological resources.[19] He suggested that a depository be established in each généralité and at each province's expense to preserve botanical and geological samples gathered by local societies of agriculture and by the engineers of the Ponts et Chaussées. From these regional collections, items could be selected for a centralized national collection in Paris. Lavoisier opposed this proposal, holding that there was nothing new in it, and proposed reviving the *Atlas*. Lavoisier showed the committee several Cassini sheets on which he had recorded geological data and a model, small-scale manuscript map for synthesizing the Cassini sheets. He resurrected his idea of 1773, to send curates, naturalists, and administrators and engineers blank maps of a region to fill in; several maps of the same region could then be compared and, finally, a critical composite map could be constituted and engraved. Demarest responded to Lavoisier's challenge by bringing in two maps of the Auvergne, one at the scale of the Cassini maps (1:86,400) and one at twice that, on which various geological formations were displayed by shadings and symbols. He proposed making several maps of the same area, each showing just one geological layer at a time, as the answer to the problem of combining cross sections with surface maps. To the charge that this would be prohibitively expensive, Demarest answered that the work could be spread out over thirty or forty years. Lavoisier then remarked that since the project Demarest proposed could be reconciled with what Guettard and Lavoisier had started two decades before, Lavoisier and Demarest should join forces. The government, after all, was unlikely to support two rival yet similar ventures. The committee was then charged with the responsibility of evaluating both proposals. To gain an advantage, Lavoisier proposed going over all of his and Guettard's manu-script material at his own expense. He further proposed that this information be compiled uniformly on Cassini maps, a task he thought would take five or six years if the government would provide him with two assistants at the cost of four thousand livres a year. (In addition, he requested that the government purchase three or four sets of the Cassini maps.) These maps would belong to the state, which would then engrave a geological/Cassini survey. The time for engraving was estimated by Lavoisier at twelve to fifteen years. The committee decided not to choose between the rivals but instead asked the government to decide—which had the effect of ensuring that nothing would happen.

National geological mapping did not begin in France until the nineteenth century. In 1811 André-Jean-Marie-François Brochant de Villiers presented a plan for a detailed geological map of France. As the work of such British geologists as William Smith and George Bellas Greenough became known in France, the incentive to catch up grew stronger in France. In the 1820s, Brochant de Villiers, Ours Pierre Dufrenoy, and Jean Elie de Beaumont were sent to England to study English geological methods.[20] Upon their return, Dufrenoy and Beaumont split the responsibility for mapping France. Using the Cassini map as a base on which to record data, they covered France between 1825 and 1836. In 1841 Beaumont presented their work to the Academy ("La carte géologique général de la France," in six sheets at a scale of 1:500,000), together with the first volume of a companion *Explication*, which remains one of the best introductions to physical geography ever written. Meanwhile, the geological cross sections of Georges Cuvier and Alexandre Brongniart (1811) to complement their map of the Paris region established a new standard that even the English adopted. These cross sections were especially interesting because the authors attempted "to indicate the degrees of certitude to be attached to different degrees of extrapolation from the observable evidence."[21] Those who resolved problems in geological mapping in the early nineteenth century take on added stature when we recall how hard it was for people only a generation before to overcome these difficulties. But the nineteenth century is not our immediate concern.

At the same time as Guettard, Monnet, Lavoisier, and Demarest were at work on geological maps of provinces and of the nation, naturalists with an interest in geology had begun to explore the Alps and the Pyrenees. With the exception of Louis-François Ramond de Carbonnières, who knew both mountain chains well, these individuals traveled in and wrote only about one mountain chain or the other.

Perhaps their concentration upon a single chain reflects the enormous personal effort absorbed by mountain travel above the verdant slopes of upland meadows and beyond the last comfortable *auberge*, into areas described on no map and in no guidebook. Arguably the greatest naturalist-explorer of mountains was Horace Bénédict de Saussure (1740–99), the first to conquer Mont Blanc (1792). Saussure was as reluctant to propose theories as Buache was eager to present them. He was preeminently a gifted observer. Yet he did not make maps. Moreover, the maps he commissioned J. L. Pictet to make for his books represented no advance on maps of mountains made earlier in the century.[22] Saussure did not experiment with graphic techniques. But others did, out of a desire to record, publicize, and compare what they learned. Three aspects of mountains suggested the need for illustration: their geological features, their height, and their complicated and variegated appearance.

One approach was to make a cross section or profile of a mountain or mountain range. Such a profile, if drawn to scale, was a scientific kind of topographic view emphasizing only the ridgeline of highest altitude. A profile could be made of an individual peak as viewed from a particular vantage point or of many peaks as a composite of several veiws. La Condamine had already published a measured representation of a topographic relief profile based upon a combination of triangulation and barometric readings. In the lower right-hand corner of his "Nouvelle carte de la Suisse," published in 1778, William Faden inserted an "outline profile—a transect through Mt. Blanc—measured with the level of Lake Geneva as base line and drawn to scale, in English feet, with notes on prominent landmarks."[23] Among these features was the snow line, that point above which snow perpetually lies. Bouguer, who had already observed the snow line in the Andes, had realized that it is not at the same elevation everywhere but descends as distance from the equator increases; La Condamine had made parallel observations concerning vegetation, linking the effects of altitude and latitude.[24] In the profiling convention, snow or vegetation lines on a mountain slope were like isometric lines, or lines of equal value.

François Pasumot (1733–1804), who served in the mapping office of the navy as a geologist, made explorations of the Pyrenees to identify the locations of minerals. Although his maps used oblique perspective crudely, they also demonstrated an interest in new kinds of information. One, for example, attempted to show the relative heights of mountains against a scale as well as the permanent snow line (fig. 14).[25] Marcel Cadet published a map of Corsica in 1789 on which the depth of water at several epochs in history was shown by means of hypothetical contour lines.[26] Jean-Louis Girard-Soulavie (1752–1813), a priest who studied his native Vivarais, had a map of that region drawn by Dupain-Triel, Sr., and made his own relief map as well.[27] Pierre Bernard Palassou (1745–1830) prepared views, cross sections, and maps of the Pyrenees, indicating mineral deposits with the spot distribution technique as a collaborator of Guettard in the 1770s.[28] The best maps of the Pyrenees were made by Louis-François Ramond de Carbonnières (1755–1827), but these were superior by virtue of the greater accuracy of their content, not because of any improvement in cartographic technique.[29]

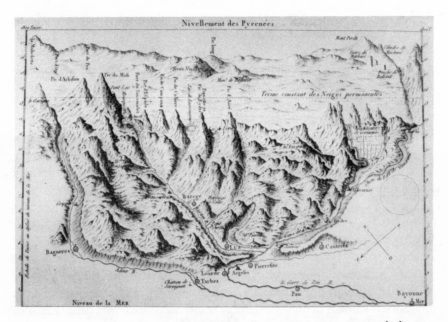

Figure 14   Elevation of the Pyrenees. This map shows the Pyrenees in perspective, with elevation correlated against the vertical scales. The dotted line shows the snow line. From François Pasumot, *Voyages physiques dans les Pyrénées en 1788 et 1789* (1797). 22 cm × 15 cm. Phot. Bibl. Nat., Paris, Ge. F. Carte 5602.

It must be remembered that those naturalists and geologists who ventured into the mountains in the first place did not do so to make better maps; they were interested in graphic techniques that would adequately illustrate their theories and the evidence they had collected. Since their purposes did not include better maps of mountains except as an incidental by-product, they should not be criticized too harshly either for having failed to make good ones or for having failed to use improved graphic techniques more aggressively.

Du Carla wrote his treatise on contour lines after many of the geologists and naturalists in eighteenth-century France with an interest in mapping mountains had already adopted their professional working habits—and even after some had done their most important work. As with ocean depths on hydrographic charts, contour lines of elevation could not usefully appear until the elevation of many places had been measured. These measurements were taken partly for their own sake, as explorers tried to discover the highest peaks, and partly for their relevance to geological and botanical studies. In 1807 Alexander von Humboldt observed that altitude had been reliably measured at only 122 sites, nearly half of these reports being his own barometric readings.[30] By 1817 altitude had been measured at 248 places in the Pyrenees alone.[31]

The only eighteenth-century naturalist to use contour lines was Charles Hutton. Hutton published calculations of the density of the earth in the *Philosophical Transactions* of 1778, adding that in the course of his work he " 'fell upon' the method of 'connecting together by a faint line all the points which were of the same relative altitude,'" thereby obtaining "'a greater number of irregular polygons lying within, and at some distance from each other, and bearing a considerable resemblance to each other.'"[32] Hutton apparently invented contour lines for himself; his idea, notwithstanding its accessibility in print, was probably forgotten. About 1800, as Sir Thomas Larcom later reported, lines of equal altitude "in the French mode" were introduced to English military cadets. He believed that these lines, called contouring, "had never been practically used in England" before. Until the 1830s and 1840s, in England, only military engineers showed an interest in contour lines.[33] Why did the English treat contour lines as a French practice about 1800? And why did military engineering provide the context for their introduction and propagation?

Mountain Maps in Military Engineering

Military engineers in eighteenth-century France were the heirs of the Marquis de Vauban, not only because that great man had given them their corporate identity, but also because he had elevated the art of fortification and had helped to define the strategic perimeter of France as a fortified frontier. This legacy conditioned the professional activities of French engineers as the custodians of the largest network of fortifications in Europe. During times of peace, military engineers were preoccupied with preparing for the next war, principally by mapping in great detail areas with which they were already familiar and mapping areas that armies might occupy or traverse for the first time. (See the discussion of planned surveys in chap. 2.) Progress in military mapping was affected by several factors. Cost-cutting bureaucrats looked upon these activities as make-work projects; engineers were themselves unsure whether their corporate identity and principal functions were primarily military or technical; and the separation of fortifications engineers (in the "corps du génie") from the "ingénieurs géographes militaires," with responsibility for reconnaissance and intelligence, created rivalries and gave rise to bureaucratic disputes that made both groups hostage to ministerial ambitions and court politics.[34] I mention these factors to remind us that military engineers were as attentive to protocol and to account books as to the purely technical aspects of their profession. Had they failed to make their way as soldiers, they would not have been given the funds or granted the mission, year after year, to undertake mapping ventures. It is worth repeating that these engineers, like scientists and mariners who entered unfamiliar and hostile environments, demonstrated considerable resourcefulness, openness to new experiences, and a high degree of professional dedication and intellectual integrity.

Military cartographers made a considerable effort to convey the shape and height of mountains realistically. There were no models for them to emulate. In the years around 1730, a cadaster had been made of Savoy on which mountains—so prominent a feature of the region—were conspicuous by their absence. Mountains were not cultivated; therefore they did not need to appear on a cadaster except as a blank area with a name, or as a symbol.[35] On the scale they worked at, the treatment of mountains by engineers was unprecedented.[36] Their maps, moreover, were accompanied by extremely detailed reports, veritable inventories of the regions surveyed, annotated with the engineers' own views on how defensive and offensive operations could be conducted. These reports were supposed to contain information on all waterways, including depth, nature of bottom and banks, places where bridges could be built and how they could be protected, and appropriate observations about mountain passes, plains, settlements, farmlands, forests, and roads. Such reports were supposed to provide the information commanders needed to decide how many troops could travel through or defend a given area and which places would give an advantage to the offense and which to the defense.[37] Maps and reports were designed to be used together.

To convey the appearance of mountains, military engineers relied upon artistic techniques—lines, colors, and shading. They even combined oblique and horizontal perspectives on the same map. L. N. Lespinasse, in his treatise on coloring military reconnaissance maps, perceived a conflict between geometric accuracy and topographic verisimilitude: the former was a function of carefully recorded measurements and calculations, the latter a function of the mapper's ability to see and draw.[38] The eighteenth-century mountain maps of military engineers were a synthesis of mapping styles. They are among the most beautifully detailed maps of that century known today (plates 1 and 2 and figs. 15, 16, 17). The manuscript maps at a scale of 1:14,000 were military secrets of such value that not even the army made copies; the printed versions were reduced to a scale of 1:86,400, so that little information of value to an enemy could have appeared on them. The best maps of mountains, therefore, had no influence on cartographic techniques used by or available to scientists in their exploration of mountain regions.[39]

Artistic techniques and secrecy were combined most conspicuously in the army's relief maps of fortifications. The practice of making a three-dimensional model of fortifications began under Louvois and Vauban. In 1697 Vauban's inventory of the collection itemized 141 pieces kept in a wing of the Louvre, where they could be seen only with the king's permission. As models deteriorated with time and fortifications were modified, some models were destroyed, others repaired, and new ones made, so that only about thirty survive from the reign of Louis XIV. Until 1743, models were built in the field and transported to Paris when finished; from 1743 to 1750, workshops were in operation in Lille and Bethune; after 1750 all work was conducted at the Ecole Royale du Génie at Mézières. In 1774 the collection was transferred from the Louvre to the Hôtel des Invalides, so that paintings could be displayed in their place. A dozen models were irreparably damaged in the move, which lasted six months and involved over one thousand trips across Paris. Models were added even during the Revolution, when the collection was opened once a year to civilians authorized by the war minister. Napoleon added more models, including one of Cherbourg and its bay over one hundred square meters in size. Napoleon captured similar models from his enemies; some French models were taken as booty in 1815. These losses were made good by French craftsmen, and the collection retained its military function until after 1870.

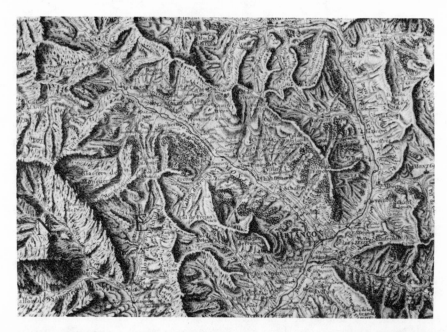

Figure 15 Area of Briançon. This shows a portion of the map of the French border with Savoy and Piedmont engraved by Guillaume Delahaye under the direction of Pierre-Joseph de Bourcet in fifteen sheets, published in 1760. This was a masterpiece of the engraver's art. Note the variety of graphic styles used to depict mountains, combining directed overhead and oblique perspectives, shading, and impressionistic contour lines. "Cartes géométriques de la délimitation de la France et de la Savoie et du Piémont." Phot. Bibl. Nat., Paris, Ge. CC. 2062.

French army relief maps were unique among those used by Europe's armies in that the same scale, one foot per hundred toises (about 1:600) was used for each. Because of their size, models had to be made in several panels, held together underneath by wooden rods. The number of panels and the placement of rods were different for each model. The top of each panel was made of thin layers of wood, which could be cut away or built up with a silk and sandscreen mesh. Military structures were reproduced from engineering plans and drawings; civilian structures were copied from notebook drawings made by engineers for that purpose. The exterior details of all structures were handled by craftsmen to show the material of construction in the actual building. From the seventeenth until the late eighteenth century, many details, such as the number of windows per house, were standardized on each model to save time; thereafter, such details were recorded with great fidelity. All this work spread among several craftsmen was coordinated by qualified engineers. They saw to it that the models were made from geodetic and topographic fieldwork.[40] The finished models were usually as beautiful as they were instructive; they provided a comprehensive, three-dimensional image of the natural, civilian, and military features of a given locality. But they were accessible only to a few staff officers in the capital.

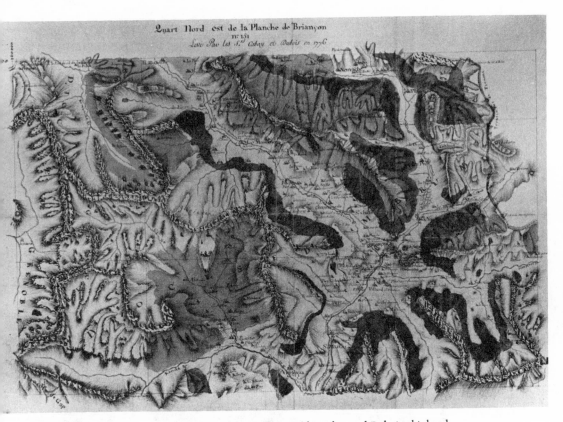

Figure 16   Cassini field survey, area of Briançon. Made in 1776 by Cabay and Dubois, this hand-colored survey showed the high elevation with thick, pointed lines and low elevation with hatch lines. Saint-Mandé, Institut Géographique Nationale.

Because the normative practices of military engineers were so satisfactory for their purposes, their incentive to innovate was a function of the few problems that did not yield easily to their techniques. Military engineers evolved innovations in the context of their routine work. One such problem was selecting a site for a fortification when this involved measuring the heights of several places relative to each other. To depict terrain more realistically, several attempts to refine hatch lines were made in the second half of the eighteenth century. In 1799 J. G. Lehmann (1765–1811) introduced a method for representing the gradient by "equivalent parallel lines," whose thickness was in proportion to the angle of the slopes.[41] This technique provided a rigid and logical framework, but it was not necessarily based on continuous and accurate measurements of an entire mountain. Contour lines were better for depicting the configuration of a small area for which many measurements of elevation had been made. Like Gaspard Monge's invention of descriptive geometry (another military secret used to design fortifications), contour lines were implicitly a graphic equivalent of measurements. Whatever their conceptual differences, contour lines

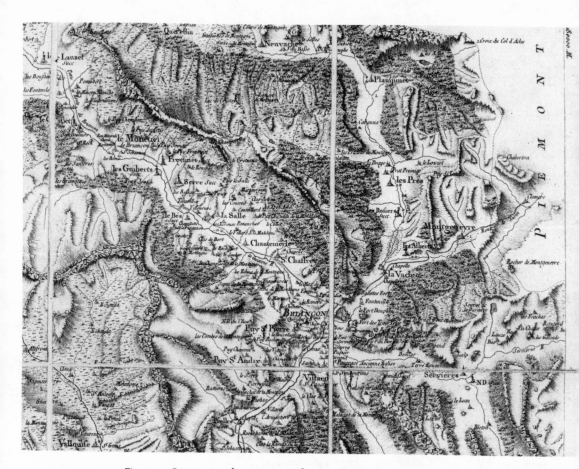

Figure 17 Cassini printed survey, area of Briançon. This is a portion of the engraved map at a scale of 1:86,400. A full sheet measures 104 cm × 73 cm. Comparison with figure 16 shows how the field survey was translated into an engraved image; comparison of Figures 16 and 17 with Figure 15 shows the superiority of the army's work. Presumably, persons consulting the Cassini surveys were never going to check their maps against the terrain in mountainous regions. From the American Geographical Society Collection, University of Wisconsin–Milwaukee Library.

and descriptive geometry shared an emphasis on representing three-dimensional space on a two-dimensional surface.

The first step toward the use of contour lines was taken in 1749 by Louis Milet de Mureau, a fortifications officer in Toulouse who was dissatisfied with the practice of making profile cross sections. The process of taking measurements and of comparing the terrain with profiles, horizontal maps, and structural designs confused him. He wanted to synthesize as much information as possible on a two-dimensional display. He proposed making observations along a traverse line that crossed the highest point of the area being mapped and then along a line perpendicular to the first one, with as many additional readings as needed. The data should then be transposed to a map, all in terms of a common level, numbered zero. Different colors could distinguish the numbers corresponding to different levels of height. The oldest surviving plan made in the field according to this method is believed to date from 1761, a plan of a fort on the island of Minorca. The highest point was given the figure of zero, the lowest point then being measured as the largest number, meaning the farthest away from zero. Such a plan was made by conceiving a horizontal plane to exist at some point above the earth. This plane was highly arbitrary and somewhat confusing, since there was no value for point zero common to different maps of different places.[42] This method was taught to students at the famous school for military engineers at Mézières in the 1760s and was adopted by the Maréchal Louis-Nicolas-Victor du Muy, who ordered its use in the making of an atlas of maps of every fortification. The idea for this atlas dates from 1771, du Muy's order from 1774. Work on the atlas was interrupted in 1778, resumed in 1783, and continued until 1789, yielding a magnificent collection of grand in-folio documents. It includes Milet de Mureau's own plan of Bonifacio, from 1786, with two horizontal planes, one above the land, the other at the surface of the sea, and readings of land elevation and water depth taken from each.

The idea of connecting all these readings for elevation into contour lines may appear to have been an easy and logical step that anyone could have made, but the fact is that despite the precedence of their use in hydrography, only one army engineer apparently understood the principle. Jean-Baptiste Meusnier de la Place was, however, no ordinary engineer. Along with some other graduates of Mézières (Borda, Charles-Augustin de Coulomb, Lazare Carnot, Prieur de la Côte d'Or, Claude Rouget de Lisle, and Monge), Meusnier appears in history books for his achievements outside military engineering.[43] Upon his arrival at Mézières, Monge had set Meusnier the "demonstration of a theorem of Euler specifying the maximum and minimum radii of curvature to certain surfaces."[44] Meusnier worked out his answers, derived from Lagrange's differential equation of minimal surfaces, in just one day. The paper in which he set out his answer more formally was read to the Academy in 1776, and the next year Meusnier became one of its correspondents. He then went on to undertake the first scientific work in aerostatics and to collaborate with Lavoisier.

In 1777—a year before Hutton's mention of contour lines appeared in the *Philosophical Transactions*—Meusnier wrote a text on the construction of a site plan.[45] In it he conceived of the contour line in terms of a fundamental problem in the design of fortifications. He began with the principle that the inside of a fort should not be visible from any point on the land around it. In his opinion, Milet de Mureau's mapping methods were inadequate because the plane corresponding to the value of zero for the purposes of measuring elevation was often far above the fort itself. Meusnier believed that the solution lay in making a map of the terrain with contour lines, which he understood to be curves connecting points of equal value of elevation. Meusnier gave a long explanation of how to construct contour lines based on their quality as tangents to a horizontal plane; in other words, he understood these lines abstractly, as aspects of spherical geometry, and practically, as denoting points that exist in space. He believed contour lines were more expressive and legible than a map on which elevations were given as numbers.

Ironically, Meusnier's first use of contour lines was on a hydrographic chart. In 1779 the Corps of Fortifications Engineers posted Meusnier to Cherbourg. In the mid-1770s, the Dépôt had sent de la Bretonnière and Méchain to take soundings along the Norman coasts in search of an anchorage for the fleet. Cherbourg was one of the sites chosen for development. During the 1780s, civil and military engineers labored to design and construct a breakwater behind which a fleet might ride out a storm. One civil engineer, Louis de Cessart, had the idea of building giant cones rising from the harbor floor, but the cones proved exceedingly difficult to construct. Meusnier was among those who expressed skepticism about de Cessart's designs. Ministers in Paris, alarmed that Cherbourg's harbor might become too costly and take too long to improve, also learned in 1787 and 1788 that there were large rocks underwater that had not been noticed before, and that might jeopardize ships. It was decided in 1789 that an exhaustive study of the harbor's bed should be made. Meusnier was in charge of this hydrographic study, which he apparently undertook with no reference to earlier work with contour lines by Buache and Du Carla.

Meusnier was especially concerned to verify the positions where soundings were made. He divided the entire harbor into triangles. Buoys were placed at regular intervals along thirteen lines constructed in these triangles. Once all the buoys had been located from the land by surveyors, sounding teams passed from buoy to buoy. Lists and tables were made indicating the soundings in the order they were made, the total number of soundings for each buoy, and the specific data observed concerning depth and the harbor bed. Maps were then made on which this information was transposed by marking the place of every sounding with color-coded symbols (fig. 18). On these maps, contour lines were drawn connecting points on the harbor bed of the same depth at one-foot intervals. Each line was color coded to indicate the nature of the harbor bed; since the composition of the harbor bed was not consistent according to depth, the color of a single contour line varied along its length.[46] Meusnier alone made the original map in order to guarantee its accuracy.

What made the Cherbourg maps important? Their scale and size, for one thing; soundings had never been made so accurately, in such great numbers and in such a difficult location before. The contour lines presented an enormous amount of information in an orderly, graphically clear, visually bold, and scientifically accurate way. More than the charts of Buache or the texts of Du Carla, Meusnier's map of Cherbourg proved the value of contour lines in practical work. Its elegance and precision justly enhance Meusnier's stature as both scientist and engineer. But having clearly grasped the practical and theoretical implications of contour lines, he did much less than Du Carla, who was far less qualified intellectually, to develop and diffuse a general model of contour lines for both sea depth and land elevation. Indeed, Meusnier's material was not published but circulated only through an informal yet restricted circle of people in government.

The secrecy surrounding military cartographic practices was lifted in 1802–03, when the army took steps to disseminate and standarize many of the techniques and procedures used by military and civilian engineers. The purpose of this effort

Figure 18   Contour map of Cherbourg harbor. A portion of the map measuring 228 cm ×
160 cm drawn by Jean-Baptiste Meusnier in 1789 at a scale of 1:7,200 as a reduction of the
original survey (1:4,320). Saint-Mandé, Institut Géographique Nationale.

was to see whether a common cartographic language could be developed for the
two groups. It can be seen as another attempt at centralizing cartography in France,
first by concentrating cartography in the hands of public officials and second by
breaking down the methodological and administrative differences that had characterized the cartographic activities of government units in the Old Regime. The
commission that met included General Nicolas-Antoine Sanson (director of the army's
map collection, the Dépôt Général de la Guerre), Baron Bacler d'Albe (Napoleon's
chief cartographer), Pierre Jacotin (at work on Napoleon's map of Egypt), Philippe
Hennequin (who had worked on the Assembly's map of departments), Prony (of the
Ponts et Chaussées), Jean-Henri Hassenfratz (inspector general of mines), Chrestien de
la Croix (head of the geographic office of the foreign service), Leroy (of the navy's
hydrographic office), and P. G. Chanlaire (of the forestry service)—in other words,
individuals with considerable experience with cartography who occupied positions
of authority sufficient to carry out the commission's recommendations.[47]

The important decisions reached by the commission involved, among other things,
standardized scales for particular map uses; the adoption of decimalization, the representative fraction for scales, and the metric system for scales and distance; a pro-
hibition against combining different projections on the same map, and in particular
against any use of semiperspective, oblique, and pictorial images of mountains (an

exception being made for coastal profiles on hydrographic charts); and standardized typefaces and lettering models. Landforms were a central concern to the commission. Too many maps contained elevation readings according to dissimilar scales because the horizontal planes from which elevations were measured were themselves at different heights. One recommendation involved distinguishing heights by color or sign according to the degree of accuracy achieved in measurement. The commission approved of contour lines for maps at a scale of between 1:1,000 and 1:10,000 if ample and comprehensive data for elevation were available. For all other situations, the commission recommended hatch lines expressing the degree of slope: the farther apart the lines, the gentler the slope, as if the lines represented the path rain would take down a hill or mountain. Colors were favored to highlight maps and were to be applied so that the land appeared to be illuminated by an oblique source of light inclined at a fifty-degree angle to the earth's surface from the northwest. The relative importance of hatch lines or contour lines mattered less in the discussions among the commissioners than whether either kind of line or coloring techniques alone could suffice. Significantly, no one referred explicitly to work by Du Carla or Dupain-Triel.

One of the first contour maps made by civilian engineers after 1803 was of Paris, and was made in Paris. Taking as point zero for elevation a level of water 1.5 meters deep in the canal basin of La Villette (chosen because the information gathered was to be used to expand the canal), Pierre-Simon Girard supervised a team that mapped Paris' topography by making gradient measurements at numerous street intersections. First they divided Paris into four zones and measured elevation along the east-west and north-south lines dividing them. Then they made detailed measurements within each zone. This work, begun in 1807, lasted several years. All readings were recorded first on the massive geodetic survey of Paris made in the 1780s by Edme Verniquet. All measurements of the same value, corresponding to an identical elevation, were connected by lines, as much as possible at one-meter intervals. A map at a scale of 1:28,000 was engraved with this information (fig. 19). Another map was also engraved using hatch lines to express elevation instead of contour lines. Two additional maps were engraved without elevation data, one showing the distribution of water from the Canal de l'Ourcq in Paris, the other showing the distribution of water from aqueducts and pumps along the Seine (each network being distinguished from all the others by lines combining different sequences of lines and dots). Girard believed these maps could be used to select the best sites for reservoirs, aqueducts, water mains, sewers, and canals.[48]

The decisions of the commission codified and popularized the best available practices. But the commission was unable to impose its decisions. Discussion on all these technical matters lasted many decades, as the use of maps and the instruments to make them continued to evolve. In 1804 there was no clear agreement on how the earth's surface should be represented and in particular on the role of maps, compared with other forms of illustration, for scientific and utilitarian purposes. As mentioned at the end of the previous chapter, in 1804 Dupain-Triel published a briefer and more

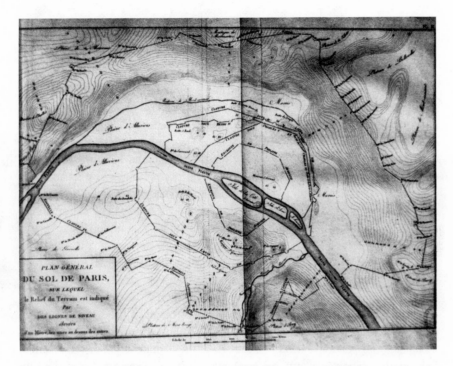

Figure 19 Contour map of Paris. This map shows the extent to which the old city, within the walls of the Middle Ages, had been settled on flat land. This was the first contour map of the city; contour lines are at one-meter intervals. from Pierre-Simon Girard, *Recherches sur les eaux publiques de Paris, les distributions successives qui en ont été faites, et les divers projets qui ont été proposés pour en augmenter le volume* (1812), plate II. 1:28,000; 34 cm × 24 cm. Phot. Bibl. Nat., Paris, Ge. FF. 4459.

technical text than his earlier books, apparently to induce Jean-Antoine Chaptal, then minister of the interior, to support the production of maps with contour lines. In 1811 S. F. Lacroix published a book in which he explained contour lines clearly and gave the chronology of their development relatively correctly from Buache to Du Carla and Dupain-Triel. But Lacroix also acknowledged that despite their potential usefulness, especially for the selection of routes for canals and roads, contour lines were still not in common use.[49] Far from marking the final acceptance of contour lines, as is often asserted, the commission's decisions represented only a major but inconclusive step in that direction.

The members of the commission were unsure whether cartography was more of an art or a science, or even whether it was in fact a kind of scientific art, with the accuracy and appeal that we would attribute today to machine-made images such as those taken by satellites, in which features of the earth are beautifully and precisely transcribed as color-coded patterns. In their attempt to depict the earth's surface more accurately, mappers were not trying to deny the beauty of the ocean's vastness or of a mountain range's craggy height. Rather, they argued that without maps we cannot appreciate the majesty and wonder of nature as well as we can with them.

From what we see on paper, we are still brought to consider why the earth has the features it has—only now we can begin to answer the question not through metaphysical speculation, but through the study of the very forms whose appearance excited our curiosity.

Mapping was a way of domesticating spaces and landforms that had formerly intimidated man and created barriers between societies. That mountains or seas could now be traversed with greater confidence itself gave rise to new initiatives to exploit them more fully. But the notion of exploitation was still free from the pejorative connotations the term would acquire after imperialism and industrialization. For example, engineers built roads and canals with generous plantings of trees and with attention to such details as a lockkeeper's house, not only for practical reasons but also to create a distinctive landscape that enhanced the region. The industrial revolution was not nearly so urban around 1800 as it became after. Claude-Nicolas Ledoux's designs for a saltworks at Arc-et-Senans and Pierre Toufaire's ironworks at Le Creusot, like Robert Owen's textile mills at New Lanark, were in remote, rural settings. By their very presence, such facilities as factories, canals, bridges, harbors, and roads aroused feelings of the sublime and became picturesque attractions for travelers, as worthy of attention as the works of nature. What late eighteenth-century Frenchmen had in mind was not so much the transformation of the seas or mountains into political and economic assets as the harmonization of these natural forms with a nation's political and economic structures. They acted as if the laws accounting for the global distribution of rock and water could be determined, such that maps of landforms could be applied to the improvement of agriculture, industry, commerce, and the military. The next chapter is devoted to maps made to apply general principles of geography and natural science to planned improvements in transportation.

Five

Transportation Planning Maps

Single-Project Maps

That better maps would be useful was self-evident to individuals and groups involved in assessing and improving France's geopolitical, economic, and social conditions. Better maps were both an end in themselves and a means to other ends. Improvements in cartographic techniques were conceptually linked to projects to change various aspects of France's spatial organization. Gradually an awareness emerged of France as a coherent territorial unit whose various parts ought to be interconnected. To document the state of France's transportation system and to improve it, maps were made, and in the process, new kinds of maps and new uses for maps were introduced.

The mapping of landforms for transportation planning was an early form of thematic mapping. As Arthur H. Robinson explains, a thematic map

> concentrates on showing the geographical occurrence and variation of a single phenomenon, or at most a very few. Instead of having as its primary function the display of the relative locations of a variety of different features, the pure thematic map focuses on the differences from place to place of one class of feature, that class being the subject or "theme" of the map. The number of possible themes is nearly unlimited. . . . An important difference between general and thematic maps and a characteristic of the latter is the portrayal of the variations within a class of features so that the pattern or structure of the distribution becomes apparent . . . [Many] maps combine functions and . . . are partly thematic and partly general. . . . A good example is provided by the display of landform data.[1]

Topographic mapping was used in late eighteenth-century France as the basis for an analytical, thematic examination of the relation between France's natural features and its transportation system.

The development of thematic cartography remained constrained by institutional, economic, and political conditions that did not always make the tasks of cartographers easier: the inability of the state to undertake all the mapping ventures it thought necessary or desirable, and its failure to undertake fiscal and administrative reforms that would have improved its mapmaking capacity; economic and administrative limitations in both the public and the private sectors; the persistence of less accurate mapping techniques after the introduction of superior methods; the irregular rhythm of cartographic innovation; and differences in the roles of engineers and scientists in generating advances in the state of the art. Yet in the final analysis, emphasis must be placed on the original ideas and projects that emerged within Old Regime and revolutionary France.

The use of maps was advocated in the belief that to look at a map was to see patterns that exist in space but are not apparent to an observer who stands in the landscape. Maps were used to *record* observations and data, depicting information about phenomena as they are distributed in space. Maps were used to *analyze* the relation between dissimilar natural features, such as mountains and river basins, and between man-made and natural phenomena more generally. Maps were used to *compare* the distribution of minerals and plants in various places, as well as differences in economic and social conditions in human populations. Maps were used to *verify* whether man-made political structures and economic activities conform to or violate the spatial organization of nature and whether man-made uses of space are rational. And maps were used to *explore* the possibilities for change as the features of urban and rural environments were rearranged on a map's surface to show how they might be at some future date; as such, maps were a tool of model building by which possible actions could be tested for their feasibility and their consequences. Whether many of the projects that involved a greater use of maps in research, analysis, and environmental planning actually effected economic or political changes is less important than the linkage created between the use of maps and the description and critical analysis of man's behavior in and use of space. Maps to improve France were made by practical men, convinced that their work represented a reasonable investment and would produce tangible benefits. Yet their work was surely visionary, for it implied that France could evolve toward an ideal spatial, geographical condition.

The image of France as a coherent territorial unit still largely contradicted everyday experience. Wherever they looked, the eighteenth-century French saw that the effects of tradition and history and of limited communication and travel on cultural, economic, and political patterns emphasized the differences between people and places. From one end of France to another, a multitude of linguistic, fiscal, and legal patterns made people acutely aware of social and territorial microunits. Topographical, social, cultural, and institutional distinctions, however, often were not neatly superimposed on each other but overlapped, giving rise to complex networks of allegiances and affiliations that sometimes generated conflict. No wonder there was an effort to redefine the institutional structure of France in purely topographical or geographical terms as a way of sweeping aside distinctions that had accumulated over time, like dead logs in a river, impeding the free flow of goods and ideas. To conceive of France as a unified, rational space was an imaginative intellectual act that defied the facts of everyday life.

The idea that France could become a unified nation-state contained contradictions that were not clearly understood in the late eighteenth century. A central concern of the integrationists was to improve transportation and communication as a means of stimulating agriculture and manufacturing. They used topographical maps to determine where changes in transportation systems should be made. But in so doing they underestimated the potential of cities to generate new kinds of goods and services, to absorb huge increases in population and wealth, and to precipitate and diffuse changes throughout entire regions. As France emerged into the industrial urban era, it did acquire coherent international boundaries and a more extensive transportation network, but it also became more visibly divided into two sectors, one much more dynamic, prosperous, better educated, and urbanized than the other. The differences that distinguished life in one part of France from life in another part changed qualitatively, but the differences did not disappear. This was to provide cartographers with an unprecedented set of challenges, which they met by enlarging the scope of thematic cartography in the nineteenth century far beyond what had been done before.

Historians now speak of "la politique d'eau" of the Old Regime, as if the government had a policy on water resources and uses, which of course it did not. Nevertheless, there were so many projects for building canals, for improving river navigation, hydraulic power, ports, and harbors, and for increasing water supplies in cities as to leave little doubt that in France the relation of water to the political and economic uses of space was of the greatest importance. People argued that France could use its water resources more effectively and that water projects were critical as catalysts in projects to advance the public welfare. Indeed, hydraulics became as important an aspect of engineering as hydrography was in geography; it is no accident that Jean-Rodolphe Perronet, one of the country's finest engineers, made his reputation in hydraulics, and that Philippe Buache was so closely associated with theories about the distribution of watersheds in the world. Whether water resources should be developed was not an issue; controversy focused on specific projects, their commercial and military implications, and their technical and financial aspects. But these controversies are not the subject of this book, which is directed toward the development of maps as a means of displaying the kind of information needed for France's water resources to be developed.

Where did rivers run? What were their depths? What areas did they drain? Where were they navigable? Answers were not easily obtained. It was difficult enough for engineers and surveyors to measure the slope of the land or to invent ways of representing land elevation and water depths on maps when dealing with a single, compact, and well-defined object such as a mountain or harbor. How much harder it became when it was desirable to collect and synthesize information about entire regions and even about all of France. That logistic and technical problems did not deter people from this enterprise speaks of their self-confidence and optimism.

Some of this interest, admittedly, was not new. Jean-Baptiste Colbert had been concerned about the condition of waterways in his efforts to improve commerce and build fleets for commerce and war, because wood for shipbuilding was floated down to the sea. A couple of maps were made in 1664 of the Canal des Deux-Mers across Languedoc before construction, and many maps were made of individual sections of the canal; but considering its scale, this canal generated remarkably little cartographic work. Most of the maps made of the canal were publicity pieces, designed to satisfy curiosity in the public at large about this wonder of civil engineering after it was finished. The army had an interest in the carrying capacity of waterways as a means of supplying fortifications and troops. An inventory of army maps made in the early 1740s listed a report on river navigation showing the limits of navigability, carrying capacity, and the average distance covered daily by a boat.[2] This sort of information, however, was not generally available, was difficult to collect, needed to be checked and corrected periodically, and was usually presented in tabular form. As France's road system was developed and as projects for canals were presented with greater frequency, the need for more systematic coverage became obvious and desirable.

The need for good maps for coordinating France's investments in transportation was clearly felt long before such maps were at hand. In 1733–34, Claude Masse made a huge (335 cm × 65 cm) three-sheet map of the Meuse, a long and sinuous river, distinguishing navigable from nonnavigable sections (fig. 20). Marginal inserts depicted villages and towns in profile as they appeared on printed maps or in documents made by military engineers, and a written report provided fuller descriptions of places of interest. It combined various graphic genres successfully and provided a realistic image of the river's path in relation to its surroundings. But this map could not be used to study the relation between the Meuse, other rivers in the area, and the region's network of roads.[3] Five maps were made in 1751 to accompany a report on the need for canals in Lorraine to connect the Moselle and the Saar, the Moselle and the Rhine, the Moselle and the Meuse, the Meuse and the Marne, and the Marne and the Saône. Finally in 1756, three projects to join Lorraine's rivers together were depicted on a single map, so that they could be analyzed. But even on this map the width of rivers and the height of elevated features were depicted on a scale different from the one used to give horizontal distance.[4]

The example of Brittany is as instructive as that of Lorraine. De Kermadec de Moustier, a nobleman and member of the Estates of Brittany, wrote a provocative, precocious plan in 1746 to make maps related to transportation. Beginning from such criticisms as that existing maps did not agree with each other, that locations were incorrect, that coasts were disfigured, and that much relevant information was omitted, the author proposed a comprehensive geodetic and topographic survey of Brittany. Because fieldwork would bring to light much information about the prov-

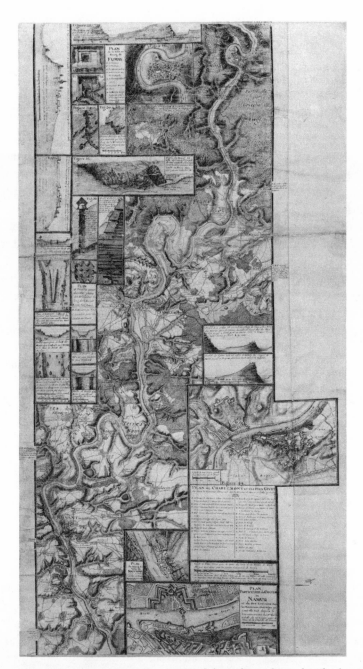

Figure 20 Map of the Meuse. This is the second of three sheets drawn by Claude Masse to show the Meuse, with inserts of striking landforms, man-made structures such as mines and fortifications, and town plans. The elegance of the lettering and the clarity of the drawing generally are remarkable. The scale is 1:36,500, and the three sheets set end to end measure 335 cm × 65 cm; they were made in 1733–34. Phot. Bibl. Nat., Paris, Rés. Ge. AA. 2053.

ince's mineral, water, and industrial resources, the mapping process would allow officials to select the best routes for roads and places for river improvements and to avoid wasteful projects. His proposal to combine a geodetic and landform survey with an inventory of resources went beyond what the second Cassini survey attempted. He planned to hire two mathematicians, a botanist, a chemist, and a specialist in commerce and manufacturing for six years and to spend no more than 180,000 livres. Particular studies, such as measurements of land elevation along the Rance and Vilaine rivers, would be undertaken at extra cost. The results of the survey could be reduced, engraved, and sold as a six-sheet map to recover the initial investment. De Kermadec's proposal was never accepted, so when a canal commission of the province wanted to evaluate a series of proposals in 1783, it had to have a set of maps made.

Canal development was supposed to link Angers, Rennes, Nantes, and Saint-Malo to provide the French with trade routes secure from interference by the British; the expansion and refortification of Saint-Malo's harbor were a part of this plan. A canal commission including Coulomb and two engineers from the Ponts et Chaussées (Antoine de Chézy and Joseph Liard) was impaneled to assess canal schemes. Maps were made to facilitate their work. Mapping involved making a geodetic base, then tracing a canal route onto the base, and finally calculating construction costs from an analysis of the condition and elevation of the land. Teams were set up to work on each proposed canal route simultaneously. Charles-Alexandre de Calonne, controller general, paid the salaries of some of the engineers, and the Ponts et Chaussées provided the services of eight students in the summer of 1784. Salaries for the project for engineers were 52,560 livres, for draftsmen, 3,837 livres, for day laborers during fieldwork, 19,003 livres; with supplies, total expenses on the maps in 1783 and 1784 were 91,639 livres. Yet in the end the maps were not needed. By the end of 1783, months before the maps were ready, the engineers on the Commission had seen enough on their tour of Brittany to conclude that technical difficulties in construction would make the canals uneconomical. The canal project was dropped until 1786, when Charles Bossut, Alexis Marie de Rochon, Condorcet, and Antoine-François de Fourcroy were asked by the Academy to examine it once again. Their recommendations—do the least possible that required major engineering work, pay attention to how canal and drainage schemes affect disease-breeding conditions, put a qualified engineer in charge who would know how to coordinate various public works projects into a unified scheme, and estimate traffic by studying settlement patterns, population growth, and economic activities in the areas served by the canals—suggested that almost all the basic research they considered essential remained to be done.[5] Clearly, there was more to building a canal or improving a river than studying topography in the field and on a map. Yet without maps, coordination of several public works projects was impossible. Good maps were necessary, but they ought to be used with other kinds of documents. Yet many were the advocates of public works projects who believed that the study of nature through maps provided the best way of choosing transportation routes (plate 3).

Plate 1 Army surveys of Briançon and environs. Part of a larger sheet (241 cm × 272 cm) made at a scale of 1:28,800 as one of a series of seventy-six sheets covering the French coast from Marseille to Nice and the land frontier thence to Grenoble, under the direction of Pierre-Joseph de Bourcet between 1748 and 1754. This particular image from 1751 displays a great variety of landforms convincingly, with color an essential medium of communication. The superiority of army surveys over those made for the Cassini map survey of France in 1776 is made apparent by comparing this image with Figures 16 and 17. Saint-Mandé, Institut Géographique National.

Plate 2   Army surveys of Montdauphin and environs. From the same survey as plate 1, Briançon, this section reveals how well army engineers represented mountains without using hatch lines or contour lines. The use of color on maps was never better in eighteenth-century France than on the manuscript field surveys of army engineers, who considered cartography a synthesis of art and science. Saint-Mandé, Institut Géographique National.

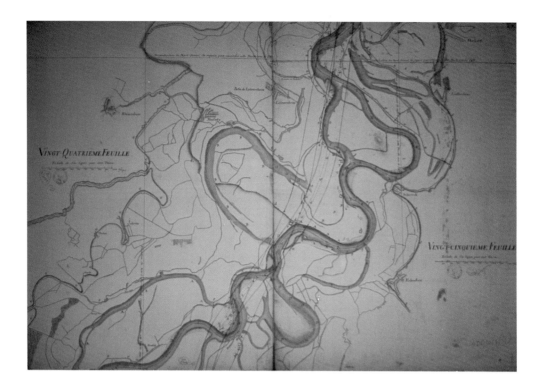

Plate 3 D'Arçon's map of the Rhine. Jean-Claude Le Michaud d'Arçon had directed survey teams of army engineers covering the Jura Mountains and Alsace (in 313 sheets between 1770 and 1787). In 1786 he was ordered to make a map of France's border along the Rhine, whose changing course obliged the French to reassess their military plans and to rebuild embankments and dikes every year. While mapping the Rhine, d'Arçon realized that the engineering works constructed each year could be shaped into a new channel broad and deep enough to contain the river forever. To display his proposal for a new permanent bed for the Rhine, d'Arçon made a set of maps in 1786 on twenty-seven sheets, each measuring approximately 49 cm × 65 cm. This plate shows two sheets of the twenty-seven positioned together to show a continuous stretch of river. D'Arçon traced a channel requiring a moderate correction to the river's channel in yellow and a channel requiring a more radical change in red; either would enable considerable amounts of land to be reclaimed. The graphic design of these maps was superb. To my knowledge this is the only instance in eighteenth-century France of someone's conceiving an impressive scheme for public works in the course of a mapping venture, as an unanticipated consequence of cartographic work. 1:345,600. Phot. Bibl. Nat., Paris, Ge. DD 5464.

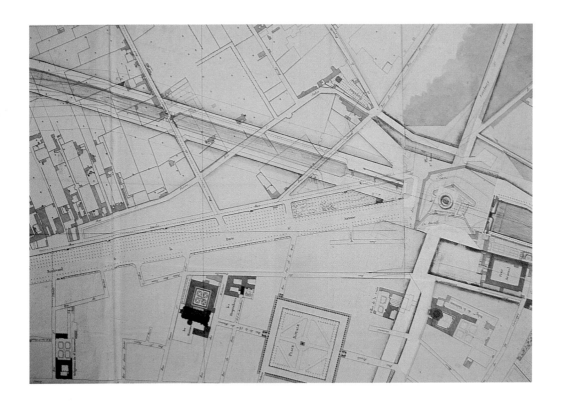

Plate 4 Plans for Canal Saint Martin in Paris. Like d'Arçon's maps of the Rhine, this sheet from a larger survey shows a proposed transportation route superimposed upon a map of the existing terrain. Unlike d'Arçon's project, this canal was actually built. The outlines in black show buildings and streets as depicted on the cadaster of Paris finished in the 1780s by Edme Verniquet. The colored lines show the proposed canal. From this map it was possible for engineers to compare different canal routes, calculate expropriation and building costs, and draw up detailed specifications for construction. The new canal cut through buildings and streets, dividing the urban fabric along the city's eastern end from the Seine to the northern suburbs in much the same way as Georges-Eugène Haussmann's famous boulevards a half-century later. 1:600; 48.0 cm × 29.5 cm. Drawn in 1811. Paris, Archives Nationales, F¹⁴ 10125-II, fol. 24.

ESSAI DE CARTE GÉOLOGIQUE ET SYNOPTIQUE DU DÉPARTEMENT DE L'OURTE, ET DES ENVIRONS.

Plate 5 Wolff's thematic map of the Ourte. This is one of the most complex and sophisticated thematic maps of the period before 1850 because of the variety of cultural, political, and economic factors that can be correlated. The thin gray line curving from the upper left side to the bottom center separates Germanic and French-speaking zones. Blue and yellow shadings denote geological zones. Printed in 1801. J. L. Wolff, "Essai de carte géologique et synoptique du Département de l'Ourte et de ses environs." 1:444,444; 50 cm × 25 cm. Phot. Bibl. Nat., Paris, Ge. D. 22101.

Plate 6 Coquebert de Montbret's agricultural map of France. On a map outline drawn by Jean-Baptiste Omalius d'Halloy and engraved in 1820 or 1821, Charles-Etienne Coquebert de Montbret depicted areas where orange trees, olive trees, and grapevines were cultivated—respectively, the orange, green, and pink areas. He used agricultural statistics collected about 1808–9. Before he made this map, he traced statistical data onto departmental maps; this map of France is a reduction of other departmental maps. "Essai d'une carte agricole de la France, des Pays-Bas et de quelques contrées voisines." 1:3,700,000; 38.0 cm × 37.5 cm. Phot. Bibl. Nat., Paris, Sg. D. 187).

Plate 7   Map of imaginary territory by a civil engineering student. One of about thirty surviving maplike drawings made as part of a final examination at the Ecole Royale des Ponts et Chaussées between 1770 and 1793, this image is typical in its size (52.5 cm × 33.5 cm) and graphic sophistication. In this undated and untitled image, the student has chosen to show two planned areas with the same spatial design—one of a rural park, the other of a city. This pattern illustrated a teaching of Abbé Marc-Antoine Laugier in the middle of the eighteenth century about the essential similarity between urban planning and landscape design. Each student had to depict a variety of man-made and natural elements. The challenge lay in arranging these into coherent and attractive patterns. Paris, Centre Pédagogique de Documentation et de Communication, Ecole Nationale des Ponts et Chaussées.

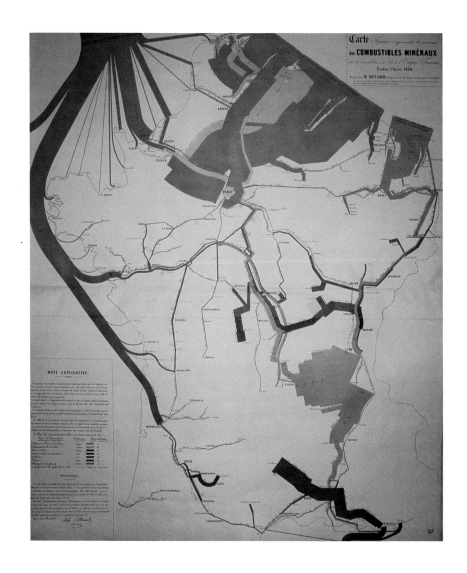

Plate 8 Minard's thematic map of mineral shipments by waterway and rail. Made by Charles Joseph Minard to show statistics for 1856 and published in 1858, this thematic map is typical of his style. The thickness of the flow lines showing the volume of minerals transported was an accurate representation of statistical proportions. Minard often distorted geographical shapes to accommodate flow lines. This map shows the impact of imports from Belgium and England, the significance of industrial development in different areas of France, and the extent of shipments between regions. The map was composed with the same attention given to graphic design in the way of coloring and lettering as would be given to a drawing of a machine or bridge. Paris, Centre Pédagogique de Documentation et de Communication, Ecole Nationale des Pont et Chaussées.

Maps of the Seine by Abbot Jean Delagrive in 1738 and Buache in the 1760s illustrate what improving maps sometimes involved. These maps were made for the prévost des marchands of Paris; the merchants were obviously interested in the condition of the river, since Paris received wood and other supplies from upstream and engaged in other trade with Rouen and, via Rouen, with the Atlantic maritime world. Interfering with river traffic were many mills, which drew power from dams and races, as well as sandbars and other natural obstacles; frequently freight had to be carried around them on land, which was costly. The Seine was also a major source of water for the capital and an important conduit of its wastes. Changes in the river's course, depth, and volume affected the city's economy. The authorities specifically wanted to know how the river could be improved to carry a heavier volume of traffic and of wood. Jean Delagrive made a series of maps, each showing a part of the river with canals, bridges, millraces, and mills conspicuously marked.[6] But each sheet, each part of the river, existed separately; there was no overall map that allowed a viewer to see how these sheets could be fitted together to form a sequential picture of the river along its length. Instead, as a summary Delagrive provided a chart of the Seine that listed all its tributaries in the order in which they merged, as a kind of genealogical chart—something far from useful. Delagrive's map had more in common with seventeenth-century maps of the Sanson style than with the Cassini or Masse maps that were being made at the time.

A generation later Buache was given the task of bringing the information on Delagrive's maps up to date.[7] He could do this only by making a whole new set of maps. His effort did not involve science per se, but he used methods of organization and principles commonly thought to be intrinsic to science. In his discussions with the authorities, Buache pointed out that in the condition they were in Delagrive's maps were useless: they lacked sequential order, geodetic orientation, and uniform scale, and they could not be fitted together. Buache then made four or five maps showing parts of the Seine above and below Paris as an example of how maps ought to look. The difference between these and Delagrive's maps convinced the authorities that Buache should make a map of the Seine between Paris and Rouen. He began by making a half-scale reduction of Delagrive's maps onto thirty-seven sheets. Then an engineer traveled along the river making corrections as necessary, during October and November of 1766 when the waters were very low, so that sandbars, submerged rocks, and the like could be examined and so that a navigable channel to be used when the river was low could be marked on the maps. To fit all the sheets together, a master map was made on the basis of the Cassini survey, supplemented by additional geodetic measurements. Then the sheets were properly oriented and placed on a uniform scale, and the points at which they could be joined were marked. A fresh set of maps was then made in seventeen sheets.[8] The river's current was indicated by arrows, and the path of river navigation and some depth readings were also marked.

The next set of maps covered the distance from Paris to Rouen as verified in 1766 in eleven sheets and from Rouen to the Atlantic in a supplemental sheet made in 1744. Another sheet provided a summary of the preceding twelve. Buache added a

"Carte générale historique et physique du Bassin terrestre de la Seine comprenant toutes les rivières dont les eaux se rendent dans ce fleuve" in seven sheets, which he had made in 1730 and on which he highlighted every locality that was engaged in trade with Paris as well as the forests from which the capital was supplied; moreover, those parts of the river on which wood could be floated and supplies carried by boat were distinguished from those parts where such movements were not possible.[9] Finally, Buache made a schematic chart of the Seine in which the river and all its tributaries were depicted as straight lines in exact proportion to their actual length, arranged as a genealogical chart might be but with a rigor and an analytical purpose totally absent from Delagrive's schematic outline, since the navigable segments of the river were clearly distinguished from the nonnavigable ones. Buache also developed a bar graph—perhaps the first of its kind for this purpose—to show the monthly variation from 1732 to 1766 in the river's height in Paris and the minimum height necessary for navigation on it.[10] A separate set of 31 navigation charts from Paris to Rouen, together with the charts enumerated above and various studies and minutes, amounted to 204 sheets. Even so, Buache did not consider his work at an end. He recognized the need for additional information, especially on the height of the river and the condition of its bed, and for periodic corrections to the maps he had already made.

Various projects were put forward between about 1770 and 1830 to build canals between loops of the Seine, to improve the river's navigational channel, and to build a canal from Paris or Rouen direct to the coast, but none stimulated maps as comprehensive and impressive as Buache's. Better maps were not made because most projects never advanced beyond the proposal stage. Buache's maps were forgotten in the meantime. One project was sponsored by Carnot in the year III for five canals between loops of the Seine, thereby reducing the distance between Paris and Rouen at a cost of 4.5 million francs; it was supposed to pay for itself in five years through tolls on increased traffic. Joseph-Marie Sganzin and Pierre Forfait, two engineers of the Ponts et Chaussées, were charged with evaluating this project. Given the shortage and high cost of foodstuffs in the year II, any proposal to improve transport was important. They reported finding Buache's map of the Seine (1763–67) in eleven sheets in the navy's hydrography office. The engineers took Buache's map on a journey from Rouen to Paris and were able to identify most of the river's contemporary features on it. By comparing Buache's map against the river after an interval of thirty years, they concluded that the river had changed very little and that a new map was not needed. Although Buache's maps of the Seine remained the best for two generations, they were not published, and so the typical map of the Seine that appeared as an illustration in a report or proposal was inferior and inadequate.[11]

The process of building canals and highways and improving harbors and rivers had profound implications for the spatial structure of cities, and thus for urban maps. Often engineers and architects viewed cities as they did rural landscapes, as areas to be reshaped into a new, more productive pattern of territorial organization. Maps based on thorough surveys were used to record and coordinate all public works projects and to calculate the value of property that might be expropriated.[12] Many projects that never advanced beyond the design stage have survived only in map form, such as the project to clear a public space in front of Notre-Dame de Paris, where a line parallel to the Paris meridian and a line perpendicular to the church's facade intersect; it was to be called the "Place du 1er Mille," because distances to Paris would be counted from it. The designer, Moreau-Desproux, thoughtfully proposed that the marble marker for distance be set in the center flush with the pavement so as not to interfere with traffic (fig. 21). One outstanding set of maps for a successful project depicted the canals running from the Seine north to Saint-Denis and the Ourcq. These maps were drawn in the 1820s on the printed outlines of the geodetic survey completed by Edme Verniquet just before the Revolution. The canal in blue and quays in brown can be easily compared with the city as it existed (plate 4). Engineers did not try to insert the canal discreetly into the existing urban tissue; rather, they designed the canal as an agent of change whose presence would create opportunities for commerce and industry along its banks. Projects for canals in Paris functioned as part of an even larger scheme to transform the capital into a commercial city attached to the ocean by improved waterways and transregional canals. Projects proposed for individual cities and for rural provinces were component parts of a national plan.

Comprehensive National Maps

Paris could afford the kind of cartographic coverage Buache provided and felt the need for it. But coverage of this kind for the country as a whole clearly called for a level of support that only the state could provide. Before the state would advance the funds and authorize the work, it had to be convinced of three things: first, that France's transportation routes could be analyzed as individual segments composing a national system; second, that the relation of transportation routes to topographical and hydrographic features was critical; and third, that good maps provided an effective way to allocate resources and to supervise the development of France's transportation system. In fact, an administrative body that operated on these assumptions, the Ponts et Chaussées, already existed. But activities were limited to royal roads and for a while to the nonmilitary aspects of harbors.[13] Military engineers thought that many of the responsibilities of the civil engineers ought to be theirs. To some extent, therefore, rivalry between the civil and military engineers may have blocked any major reallocation of responsibilities and funds for engineering work and would have kept the government from undertaking a systematic mapping effort for transportation planning. There was in fact no single body with jurisdiction over France's land and water routes, capable of coordinating the roads, canals, and harbor developments launched privately and by the military and the civil authorities. Nevertheless, that the road and waterway elements of France's transportation network should somehow be studied together as if they formed a system, and that the way to do this was with maps, were attractive ideas in late eighteenth-century France. Before considering a variety of proposals for the initiation of mapping ventures to combine topographical

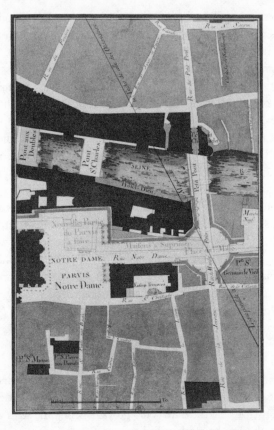

Figure 21 Plan for a Place du 1ᵉʳ Mille, Paris. Drawn by Moreau-Desproux (1769), this image shows the outlines of buildings to be removed from in front of Notre-Dame and of a new ensemble to be built as far as an intersection aligned with the Petit Pont. The center of that intersection would be the point from which distances from Paris to other places in France and the world would be measured. The oblique dark line crossing that intersection was drawn parallel to the Paris meridian. From the author's "Plan générale des différents projets d'embellissements les plus utiles et les plus convenables à la commodité des citoyens et à la décoration de la ville de Paris." Phot. Bibl. Nat., Paris, Département des Estampes, petit-folio Ve. 36.

features with transportation structures, let us first consider the kind of maps the Ponts et Chaussées engineers made and the uses of these maps in decision making.

Thanks to the pioneering work of Guy Arbellot, the maps made by the Ponts et Chaussées for its own needs—to design, construct, and maintain bridges and highways—have been lifted out of the shadows and dust of the archives. The origins of the Ponts et Chaussées strategy in the context of the first Cassini survey of 1733–44, discussed in chapter 1, clearly involved centralizing control of the Ponts et Chaussées in the mapmaking and map-collecting functions of a drafting office in Paris. In 1738 civil engineers had been ordered to make detailed highway maps of the area for which they were responsible at a scale of 1:8,640. Each sheet of paper (75 cm × 31 cm) represented a length of road of about 3,323 toises (or 6,480 m). The engineer in the field was supposed to work from a geodetic matrix based on Cassini's calculations and his first national survey. He was supposed to add such details as groups of houses, gardens, courtyards, streams, woods, vineyards, mountains, and civil boundaries up to a distance of 600 toises (or about 1,170 m) on each side of the highway, so long as such details could be depicted clearly on the map. Places of interest farther from the road but visible from it were also to be shown. Fieldwork was to be accomplished with a compass and a graphometer or half-circle and with chains of ten toises. Colors to show different road-building materials were also specified.[14] The engineers in the provinces were to make a copy of each map and send the original to Paris.

Perronet, for his part, recorded the progress of road building in twenty-two généralités in a register, noting in tabular form a road's length, the date of its map, the name of the mapper, whether the final plan was made in Paris or in the provinces, and the locations of various relevant memoranda and drawings.[15] At the beginning of 1776 Perronet counted 2,090 maps and 757 drawings of bridges in his collection. From this documentation he estimated the length of the road network under his control at 3,135 leagues (about 12,932 km); roads in the pays d'élection, under direct provincial control, amounted to perhaps a quarter or a third more.

The maps in Perronet's collection were easily reduced (to 1:17,280) into more compact formats. Three collections of these reductions have been found: one for the king, one for the administrators of the Ponts et Chaussées, and one by a Benedictine geographer, Dom Guillaume Coutans, who may have intended to publish it.[16] Each compact edition contained only a selection of the material sent to Paris. Fourteen atlases were drawn between 1752 and 1771 for the king; twelve depicted complete itineraries from Paris to a point on the frontier in a remote province. Each page in the atlas made in 1762 of the road from Paris to Lille cost two-hundred livres to produce from fresh field surveys; a single atlas covering only a modest distance therefore cost a few thousand livres.[17] The atlases made for the engineers' own administration duplicated the format of the royal atlases but were not drawn and finished with the same artistic care. The third set implemented a suggestion made in 1767 to engrave maps; Dom Coutans published a small atlas of twenty-two pages in 1775 that accompanied written descriptions of the road from Paris to Reims.

Two printers, Michel and Desnos, published maps and timetables for the use of highway travelers and shippers that condense much information about roads in a schematic format. Their first *Indicateur fidèle, ou Guide des voyageurs* of 1764 was reedited several times. It presented the road network in simplified form at a scale of 1:200,000. The type of coach service, the day and hour of its departure from Paris, the places and times of stops, and the distances between stops were noted on the map and in a table in the margin. From successive editions of the *Indicateur fidèle*, increases in speed and improvements in road conditions can be calculated.[18] To increase speed and let carriages traveling in opposite directions pass easily, engineers widened existing roads and built new ones. They also eliminated curves as much as possible, designed bypass routes around villages and towns, and investigated the properties of various road-surfacing materials.[19] The most important efforts had been made on roads linking France's frontiers and provincial centers to Paris, with secondary emphasis on networks linking France's major provincial centers to each other. Astounding gains in speed were registered. In 1765 Angers was six and a half days from Paris at a daily average of forty-nine kilometers; in 1780 it was only three days away. Five days were cut from a trip between Paris and Rennes, seven days from the Paris–Strasbourg run, and five days from a trip between Paris and La Rochelle.

Paris's sphere of influence was enormously enlarged in area as the duration of travel was reduced. This especially affected places three-hundred to six-hundred kilometers away, which fell more strongly into the capital's orbit. Large provincial cities also benefited from more rapid communications not only with Paris but also with each other. Insofar as the coach and road network was the most rapid path of communication, it is significant to observe this shortening of temporal distances between places in France immediately before the Revolution. The Massif Central and Brittany, however, were not noticeably closer to Paris in time; and one can wonder how well integrated into this transportation communication network were France's small towns and villages near the great roads. But the overall impact of these maps is nonetheless impressive: they recorded dynamic, multifaceted changes that were the civilian equivalent of those maps so dear to the general staff, depicting France as a series of garrisons connected by roads that were colored or coded to show how many troops could move between them in a day. The effect of faster, better travel on war involved everything from the organization of supplies and the location of forts to the formulation of strategies based upon rapid movement and communication. The effect on civilian activities was no less profound for being less obvious. Speed and predictability—until a generation before scarcely to be mentioned in connection with overland travel—had become normal.

The example of the Ponts et Chaussées, while important in itself, would leave the wrong impression if the reader thought the civil engineers were completely satisfied with the maps they used. An unsigned, undated proposal in the archives of the Ponts et Chaussées, probably from the 1780s, takes us back to waterways. It was entitled "Projet d'une carte de France, réduite au tiers de l'échelle de celle de l'Académie, destinée pour l'Administration des Ponts et Chaussées."[20] The "carte de l'Académie" referred to the Cassini 180-sheet survey at a scale of 1:86,400. The author of this proposal considered the maps of this Cassini survey unsuitable for administrative use because they covered too small an area for them to be arranged so that major public works projects could be designed and monitored on them. By contrast, the one-sheet summary map made by the Cassinis eliminated too many details and simplified the road network. The older Cassini survey from the 1730s and 1740s was totally out of date, and even many of the sheets of the second, newer survey had been drawn before many roads were built or improved. And neither Cassini survey had paid close attention to landforms.

The author wanted maps to analyze the relation between waterways and roads. He even made a model of a map on which existing and proposed civil engineering works for both roads and waterways could appear together. At a scale one-third that of the large Cassini survey, for example, 1:259,200, one sheet would be equivalent to nine sheets of the Cassini 180-sheet survey. (The author allowed for the possibility of a reduction to one-fourth, or a scale of 1:345,600, but preferred one at one-third.) Each sheet would include all the settlements on the larger Cassini maps but only those châteaus that were directly alongside main highways; all roads projected or built; all canals, rivers, streams, and swamps, on the same scale as the roads, with attention to navigable and nonnavigable parts; all mines, forges, glassworks, paper-works, factories, and the like; woods and forests with their area in arpents noted so that comparisons could be made between them and cumulative totals; and finally mountains, which were only sketched in on the Cassini maps. These maps should be made so that, as much as possible, the lands of a généralité were not divided up onto many different sheets. The model demonstrated italic and roman typefaces of two sizes, to see which was best; the author preferred roman for important place-names and italic for smaller localities.

Louis Capitaine, one of the principal collaborators of the Cassinis in their survey during the 1780s, began work on a reduction of the Cassini maps to one-fourth, one sheet containing the area covered on sixteen Cassini sheets. His map would have met the needs of the engineers. It was presented to the Constituent Assembly in 1790 but was not engraved until 1793. Meanwhile, on 1 July 1792, instructions were sent to the chief engineer in each department to record roads built and projected as well as canals and waterways and proposed improvements to them.[21]

What was important about the model for a reduction and its accompanying proposal, and about the 1792 instructions and the Capitaine effort, was the desire to relate water and overland transportation networks to each other and to human and natural geography as accurately as possible. The older Cassini surveys of 1733–45, which had provided the geodetic foundation of all subsequent mapping in the Ponts et Chaussées, were no longer of much use. The second Cassini survey was not suitable to the administration's purposes. A new map was needed. The danger to be avoided was that information about roads and waterways, and about human and natural geography, might be produced and mapped separately, in ways that inhibited analysis of their interrelatedness. Other cartographers in and outside government service shared similar objectives and concerns in the late eighteenth century. The Ponts et Chaussées proposal was unique in its specifics but not in its overall objective.

Proposals for maps combining topography, roads, and waterways were characteristic of the reform-minded late Enlightenment. If their authors had any doubts about the merits of their proposals or the means of executing them, these were not mentioned. Some maps appeared as illustrations in French agronomy texts or in works advocating agricultural reform.[22] In 1762 Henri de Goyon de la Plombaine published *La France agricole et marchande* to show how various regions could be cultivated more intensely and profitably.[23] It contained three maps: one of part of Champagne was coded with letters corresponding to parts of the text where techniques and proposals for specific places were discussed; another showed areas drained by the Dordogne River; a third showed where canals could be build in the Cevennes. These maps were subordinate to the text; they functioned like illustrations in a book. The kind of map reformers often had in mind was different. They wanted fresh surveys of France as the basis for analysis and planning.

In presenting and evaluating suggestions for mapping ventures, people followed certain procedures. Individuals in government service proposed projects to their superiors, who passed judgement on them and often submitted them to others with a reputation for expertise, inside or outside the government. Thus when Pierre Cornuau (a surveyor whose work on cadastral surveys in the Limousin was mentioned in chapter 2) wrote to the controller general from Limoges to propose that a map be made of land elevation from Mont-Dore to the sea, his suggestion was sent to Condorcet. Condorcet found that Cornuau's proposal contained inaccurate knowledge about the shape of the earth and about barometers, but nevertheless he thought the project had merit. He suggested that an outline map could be given engineers from the Ponts et Chaussées to fill in with as much information as possible about rivers, mountains, and roads.[24] Cartographic proposals were also presented by people outside the government, either to someone with influence in the government or to the public at large by publishing a book or pamphlet to attract attention and support.

Military personnel submitted several proposals, of which the following is representative. In 1777 mapping of Provence and Dauphiné brought forth a statement on the usefulness of a map based upon measurement of land elevation and water depth, and of the width, depth, and place of highest elevation of waterways, in Flanders, Artois, and Hainaut. In the absence of such maps, it was observed, hydraulic projects might be built in several locations that would jeopardize each other, since the flow and volume of water could be too great or not enough.[25] A map was in fact made of Flanders and Artois containing a slightly different selection of information: roads, streams, and canals, already built or proposed, at a scale of 1:177,680, with remarks on the size and carrying capacity of canal boats and on whether they were pulled by men or by horses.[26]

In 1777 an official in Paris sent the intendant of Languedoc a map on which he was to trace all bridges and roads, constructed or projected. The intendant was told that this map would be used to make a "Carte générale des routes et chaussées du Royaume." The intendant returned the map, completed, to Paris, but neither it nor any other trace of this proposal of 1777 has been located.[27]

Jean-Antoine Fabre first presented his ideas on waterways in mountainous regions to the Academy in 1780. His objectives were to deepen rivers for navigation and reforest mountain slopes to limit runoff and flooding. In an undated book (probably published in the 1790s) Fabre offered the contour line as his own invention and suggested that the state use contour lines to make accurate, detailed maps of all the roads, canals, and waterways whose upkeep was a public charge.[28] Each map should use the same symbols and should be made to the same scale; sea level, taken at its lowest level in Toulon, should provide the common denominator for contour lines. Thus any information on one map would be comparable to that on any other. Fabre thought that the Revolution, by abolishing the Old Regime's domestic political boundaries, at last made possible the study of France's surface with sole regard for its natural features. But in this, as in his supposed invention of contour lines, he had been preempted by someone else.

The first person to design a mapping venture that focused on France's topography and transportation infrastructure but totally disregarded its civil boundaries was probably Etienne Claude Baron de Marivetz. De Marivetz brought his ideas to the attention of the public in books published in the 1780s. In 1779 he began publishing a fourteen-volume work entitled *Physique du monde*, of which only six volumes ever appeared. In this publication he announced his intention to print a "Carte générale, physique et hydrographique" of France in several sheets, together with a two-color summary map on which coverage of these several sheets would be correlated with the divisions of the second Cassini survey. In addition, he intended to publish studies of canals under development and illustrations of hydraulic works. Commencement of publication was announced for March 1780, and installments were supposed to be produced at six-month intervals for three years. The complete set of maps and illustrations was to cost 360 livres, and a list of engravers, printsellers, and booksellers in France and in foreign countries was published. Obviously, interested persons were supposed to initiate subscriptions through these outlets; de Marivetz promised that a list of subscribers would eventually be published.[29] Apparently de Marivetz had considerably overestimated his market. Whether anyone subscribed is impossible to determine, but the number must have been very small.

In 1786 de Marivetz published another "Prospectus à la seconde partie de la physique du monde ou à la carte hydrographique de la France," with an additional work, a general treatise on inland navigation throughout the world. De Marivetz claimed that his maps could be applied to specific tasks of public administration such as determining where canals should be built and coordinating inland navigation with the road network. His thinking was grounded in the widespread but occasionally controversial view that the construction of canals could not be left to local or private initiative but had to be planned by a central authority. De Marivetz promised to present all possible routes for canals so that the state could choose which ones to develop.[30]

De Marivetz also intended to make twelve maps corresponding to twelve epochs in the history of France, with those parts covered by water in each epoch marked in blue. De Marivetz was a neptunian in his interpretation of natural history. Presumably he wanted to show the natural tendency of river formation and land drainage.

A map devoted to the present was to appear in forty-five sheets. This map would belong to a new kind of physical geography, in which natural phenomena alone matter. As a complement to the map, de Marivetz proposed a "dictionnaire des noms des terreins," featuring topographical phenomena classified according to their shape, size, and relative height. The map itself was to display all the waterways of France, together with the gradients, configurations, and elevations of land above sea level and the major roads. Geological information gleaned from Etienne Guettard's mineralogical maps was to be added. The sheets were to be designed to permit assembly into a twelve-foot square. No civil boundaries were to appear, not only because such lines would clutter the map, but because the map was intended to display natural phenomena and transportation routes exclusively. The crests of plains and of mountain ranges were to be colored red, naturally navigable waterways blue, improved waterways and canals green, and projected improvements yellow. The rest of the map was to be shaded black. An assembled forty-five-sheet map would permit the evaluation of all projects submitted for rivers, ports, canals and roads, the selection of specific projects, and the integration of these projects into a completed transportation infrastructure to proceed cartographically. De Marivetz further proposed that engineers of the Ponts et Chaussées should make maps of the Seine, Loire, Garonne, Rhine, and Rhône river basins at a scale of about 1:20,000 showing all streams, bridges, paths, and roads with as many measurements of elevation and slope as possible, but with no settlements shown. Several sheets of each river basin would form a comprehensive cartographic record of its progress from its source to the sea.[31]

Finally, in 1788, he published a text with a map. The text was entitled *Système général, physique et économique des navigations naturelles et artificielles de l'intérieur de la France et de leur coordination avec les routes de terre; première partie.*[32] It contained a study of existing canals and an enumeration of projected ones, and it made an interesting and useful supposition about the economics of transport of the kind that would become common in the nineteenth century: de Marivetz estimated that each section of a canal one league long serviced an area up to four leagues inland on each side, measured as the distance loaded wagons could travel and return in a day. The map's title was equally expressive of its content: "Carte physique et hydrographique de la France, ou Carte figurative des navigations naturelles de ce Royaume, des navigations artificielles déjà existantes de celles dont l'establissement est ordonné et de celles qui sont désirables pour compléter le système général de la navigation intérieur." It was an abridged, modest version of what de Marivetz had so ardently desired to execute during the preceding decade, limited to a representation of the principal waterways, existing and projected canals, and the ridgelines of land masses separating major watersheds. The river basins were marked in yellow; places where de Marivetz recommended improvements were in red. De Marivetz had still not abandoned his intention of eventually completing a general treatise on physical geography, part text and part map, in which general principles would be applied to particular land areas and to the design of a transportation network appropriate to them.

Other people were at work at the same time on general survey maps of France's topographical features related to transportation projects. In the 1780s Emiland-Marie Gauthey, director of canals for the Estates of Burgundy, prepared a map of watersheds in France outlined as caterpillarlike darkened ribbons reminiscent of Buache. On a blank area of the sheet, off France's northwest coast, Gauthey drew several approximations of cross-section profiles of France showing changes in land elevation along imaginary lines from east to west. Nicolas Fer de la Nouerre, author of *Science des canaux navigables* (1786), brought out a map of France's river networks on which he

Figure 22 Map for improving inland navigation, Fer de la Nouerre. Engraved from material by Nicolas Fer de la Nouerre in 1787, this map was accompanied by a legend explaining its symbols. Shown here is a portion of the map showing proposed connections between Paris and western France and between Paris and the Channel. "Carte élémentaire de la navigation du royaume." 1:2,222,220; 70.5 cm × 47.0 cm. Phot. Bibl. Nat., Paris, Ge. C. 1269.

used conventional symbols (anchors, lines, etc.) to show the limits of navigability, proposed canals, and stretches of river to be improved (fig. 22).[33] Fer de la Nouerre used symbols of his own invention to differentiate between several projects. Straight lines indicated proposed highways. Stars interrupted these lines at points separating watershed basins, the number of stars indicating the degree of difficulty anticipated in executing a project. Two parallel lines indicated an existing but inadequate water route; three parallel lines, a connection adequate to the existing volume of traffic.

Dupain-Triel, Sr., was familiar with the concept of contour lines from having been Du Carla's publisher. However, his "Carte générale des fleuves, des rivières et des principaux ruisseaux de France avec les canaux existants ou même projettés" (1781) did not yet use contour lines.[34] Perhaps he was concerned that his intended readers would not be able to interpret them; perhaps he had not yet translated enough information about river basins into this new graphic form. The first contour map of France, made by Dupain-Triel, Jr., in 1791, was too small for engineers or planners; it was rather a prototype, a demonstration of the feasibility of contour lines. In 1799 Dupain-Triel, Jr., brought out a second edition of the contour map of 1791. In the margin of the 1791 contour map, Dupain-Triel had included a model cross section of land elevation across France along a line extending between the Rhône in the east, across the Auvergne, to the Gironde in the west. In the 1799 map Dupain-Triel substituted a table ranking the elevation above sea level of some forty places in France.

Dupain-Triel, Jr., made another map in 1793 that resembled his father's map of 1781 and was designed to complement a statistical abstract of France's waterways as a navigational system that he also published in 1793 (fig. 23).[35] The map of 1793, made without contour lines, grouped waterways into systems or watersheds. Mountains, which did not appear on the 1781 map, were added to the map of 1793. Places on this map were marked with symbols to indicate the limits of navigability on a waterway, where improvements could be made, and the like. The text accompanying the map was a "dictionnaire hydrographique," a listing of waterways in France and Belgium divided by department. The information had been gathered by an engineer of the Ponts et Chaussées named Bournial in 1781 and 1782. It specified for each waterway the farthest locations to which boats could travel inland; the waterway's outlet; its length; whether it was navigable seasonally or all year; and the size of the largest and smallest boat that could navigate on it, by outer dimensions and by weight.

In 1796 Dupain-Triel published an *Essai sur les moyens d'arriver à une hydrographie complète de l'intérieure de la République*, which carried the map and table of 1793 further. Acknowledging his respect for civilian and military engineers, Dupain-Triel presented his proposal as the culmination of a tradition extending back to Vauban.[36] Dupain-Triel stated that the maps he had made in 1791 and 1793 were based on inadequate data. In his essay of 1796 he called for a systematic data-gathering operation. Elevation should be recorded along the axes of the Cassini surveys, the grids of triangles from Dunkerque to Perpignan and from Strasbourg to Brest. The base, or point zero, should be derived from the average height of the Seine in Paris. At principal places along roads across France, elevation should be determined and markers with the data recorded on them should be erected. Engineers from the mining corps, Dupain-Triel urged, should be sent out to do these things. Each department would eventually get a map of its area at a larger scale; these in turn would be reduced and combined to form a new contour map of all France. Obviously, as Dupain-Triel's plans evolved, they became more complex and called for greater investments. Dupain-Triel believed these difficulties could be surmounted by a government devoted to the glory of the nation and the prosperity of the people. But the publication in 1804 of a work by Dupain-Triel recapitulating the theory of contour lines and advocating their application to topographical mapping hints at the magnitude of these difficulties and poignantly exposes the author's frustration.

Unlike de Marivetz, the Dupain-Triels did not link their maps to the production of a systematic theory about the structure of the earth. They were more aware than he of how inadequate was the statistical information about landforms based on accurate measurement. Progress in collecting more and better data was slow. In his instructions to the army's engineers in the year X, General Sanson wrote that measurements of elevation were being wrongfully neglected and that insufficient attention was being paid to the study of the relation of waterways to landforms.[37]

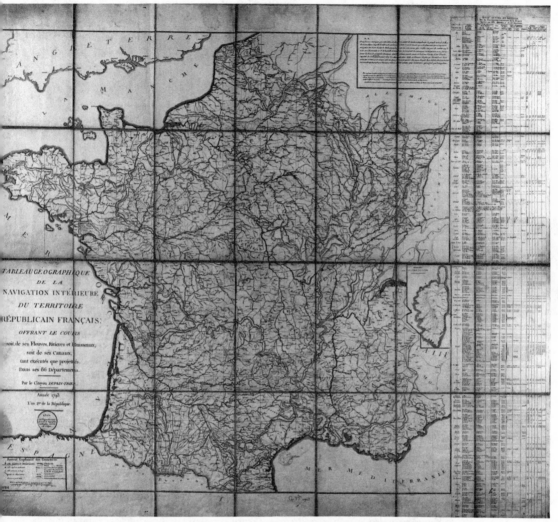

Figure 23   Map of inland navigation, Dupain-Triel. Engraved in 1793, this map shows ribbonlike lines similar to those employed by Buache, and used to the same effect: to demarcate watersheds along lines of elevation. The table along the left presents statistics about the carrying capacity of rivers and canals. Jean-Louis Dupain-Triel, "Tableau géographique de la navigation intérieure de territoire républicain français, offrant le cours soit, de ses fleuves, rivières et ruisseaux, soit de ses canaux, tant exécutés que projettés dans ses 86 départements." 114 cm × 94 cm. Phot. Bibl. Nat., Paris, Ge. FF. 10963.

In this case the right kind of maps were conceived a generation before enough data existed for their fabrication.

Proposals for maps relating natural features, waterways, and roads are important above all for the assumptions they were based on. Their authors were willing to rely upon skilled and trained experts to collect valid information in the field, but they did not ask why the necessary sums of money were not forthcoming from either the private or the public sector to advance their ventures. They believed that with information of higher quality and in greater quantity, rational decisions could be reached objectively, free from political pressures and self-interest. They thought maps were very effective instruments to collect, record, and interpret information about the environment and its uses and they assumed that with good maps, wasteful mistakes could easily be avoided. They did not entertain serious doubts about a state's ability to use maps effectively. During the Revolution, Nicolas François de Neufchâteau, as minister of the interior, eager to create "un système général de navigation intérieure qui embrasse, dans ses combinaisons, l'universalité de la France," called for the creation of twelve commissions of scientists, engineers, landowners, manufacturers, and merchants to make field surveys and to examine cartographically such major inland waterways and canal routes as the Rhine-Rhône. The list of rivers and canal routes the minister wanted examined grew longer, but the commissions produced nothing.[38] Pierre-Louis Dupuis-Torcy and Mathurin-Jacques Brisson no doubt believed uncritically that they were advocating a rational procedure when they wrote in 1808 that the proper route for a canal cannot be determined by studying the land on a field trip, but only by studying a map of the land.[39]

Not everyone, of course, believed maps were indispensable. When Turgot became controller general, major problems had arisen in connection with three canals under construction in Picardy and one in Burgundy. Turgot turned to Condorcet for advice. Condorcet counseled: "Beware also of those who, seeing two rivers separated on the map of France by a little scrap of white paper, propose to join these rivers and call that a project."[40] In 1775 Condorcet suggested that a commission composed of himself, Jean d'Alembert, and Abbot Bossut be established to evaluate inland navigation. Bossut then performed a series of experiments on the problem of the resistance of liquids to bodies passing through them (published as Nouvelles expériences sur la résistance des fluides in 1777), from which the conclusion was drawn that the tunnel proposed for the Picardy canal was not feasible. At the same time, Bossut was appointed to a new chair of hydrodynamics at the Louvre to give instruction and to conduct research relevant to rivers, canals, and mills. (To be sure, there was friction between Bossut and rival centers for the study of hydrodynamics, the Ponts et Chaussées under Perronet, and various naval programs). Condorcet, meanwhile, conducted a study of the Somme valley to see the conditions of life in marshy areas. Concerned that water would seep from canals, Condorcet found that the average expectation of life for people in marshy areas was 20 percent less than that for those living in dry areas and urged Turgot to stop all work on both the Picardy canals and the Canal of Burgundy until these projects could be evaluated more scientifically.[41] The methods of Condorcet and Turgot were thorough and multidisciplinary; they suggested that there were serious limitations to the map-oriented perspectives others adopted uncritically. (The evaluation of the Breton canal mentioned earlier expressed a similar point of view.)

In Britain, in contrast with France, the development of roads did not stimulate cartography, first because surveys were usually made and paid for by a turnpike trust or landowner, and second because few road projects involved land acquisition. As a result, road maps were often made by part-time, amateur surveyors. Canal building eventually had a greater impact on cartography in Britain, since land acquisition and disputes over alternate routes were resolved more easily with maps. Nevertheless, the use of maps was slow in coming.

Maps were made, if at all, only after field inspections had resulted in the selection of an acceptable route and construction had begun. In 1774 the House of Commons "passed a Standing Order requiring a map of the intended canal to be deposited at the same time as the bill, accompanied by a book of reference giving the names of the owners and occupiers of all lands to be crossed, and indicating whether or not they assented to the project." As of 1791, the map had to be deposited before 11 November prior to the application to Parliament for a canal act; and from 1814 it had to be of a scale not less than three inches to the mile (about 1:21,000).[42] This scale was not large enough to accommodate anything more than a limited selection of detail. As any inspection of British archives bears out, maps for canal projects were only illustrations of information, not planning documents providing the basis for a critical analysis of the project. There was a shortage of qualified civil engineers, and those few at the top of the profession (John Smeaton, John Rennie, Thomas Telford) who traveled back and forth across England and Scotland usually delegated mapping to qualified local surveyors. No comprehensive maps of England's topography and waterways were made during the great age of eighteenth-century improvements on which to plan or to record progress in constructing roads, canals, and river improvements regionally or nationally. Yet the success of the English in developing their infrastructure suggests that the lack of such maps did not create unmanageable difficulties. By 1830 about two-thousand miles of canals had been constructed. To my knowledge, no one in Britain expressed the need for the kind of highly synthetic topographical map that was so clearly desired in France. Nor to my knowledge can a single cartographic innovation be traced to the great age of British canal building, with the exception of Smith's geological maps of stratification.

By contrast, the advocates of maps in France saw nature as a valid template for human activity. They assumed that to the extent that economic activities and developments conformed to France's topography, France would become more prosperous. The political corollary of this concept was that political boundaries based upon natural divisions were in fact natural, and hence logical, rational, and stable. Projects for the development of France's transportation infrastructure reflected unquestioned bias in favor of natural features as the only rational foundation. The same engineers who contemplated clearing and rebuilding urban districts looked at low-lying lands with drainage schemes in mind, studied hills with an eye to routes for canals and roads, and searched the seacoast for harbor sites. Economic activities would follow in the builder's wake, engineers confidently believed. They accorded a higher value to superior transport than did many businessmen. In France, topographical and thematic cartograpy became conceptually indistinguishable: a topographical map was looked on as a thematic map, because the principal criterion for analyzing man-made patterns was conformity to nature.

Six

Thematic Cartography, circa
1790–1850

Early Thematic Maps, to 1820

During the second half of the eighteenth century new uses were made of maps that considerably enlarged the analytic power of cartography. A change in scale, in cartographic symbols, and in the selection of information represented was enough to convert a general map into a partly thematic one.[1] Arthur H. Robinson has stated that "the pure thematic map focuses on the differences from place to place of one class of feature, that class being the subject or 'theme' of the map." Thematic maps may be devoted to an aspect of the environment such as the distribution of plants or geological formations, but they are best known for presenting information about social or economic activities, "to discover the geographical structure of the subject."[2] Thematic maps were developed to convey specialized information and to facilitate the analysis of spatial variables in nature and society. The development of thematic mapping appears to have generalized and simplified map usage. Thematic maps heightened awareness of the extent to which spatial analysis reveals patterns in nature and society. But the emphasis thematic maps gave to spatial conceptualization aroused some controversy. Thematic mapping in France and elsewhere developed fitfully. Its potential was clearly grasped by the 1790s but not realized until a generation later. The roles of science and engineering in its formative period highlight changes in the institutional and intellectual framework for innovative cartographic activity in France.

Isolated thematic maps can be found throughout the eighteenth century, but insofar as maps are still being discovered in libraries and archives, statements about the oldest map of a certain type or about use of a certain symbol must be tentative. The oldest thematic maps were probably designed by persons who were not conscious of the characteristics of this cartographic genre. Their use of symbols illustrates the limited suitability of traditional techniques in thematic maps. For example, small-scale maps of regions and countries frequently used symbols to indicate the presence of educational, ecclesiastical, legal, and political institutions in cities. An old and influential city with several such institutions would have several symbols next to its name on the map, and at some point the level of information would become unmanageable. The user can scan the map to find all university cities, for example, but to decode all the symbols for a single city one must shift back and forth between the legend and a cluster of symbols. One undated map that is among the oldest economic maps, "La France commerçante," tried to convey information about production in brief phrases written on the appropriate area of the map (fig. 24).[3] Both techniques—brief phrases and symbols clusters—are useless for making comparisons within a category. Their function in eighteenth-century culture was to combine some of the features of a general map—the location of cities and regions—with the kind of information found in tables or charts.

One of the most curious and original early thematic maps presented anthropological information about the degree of civilization of various peoples around the world.[4] Like most thematic maps, this one intended to show a series of relationships

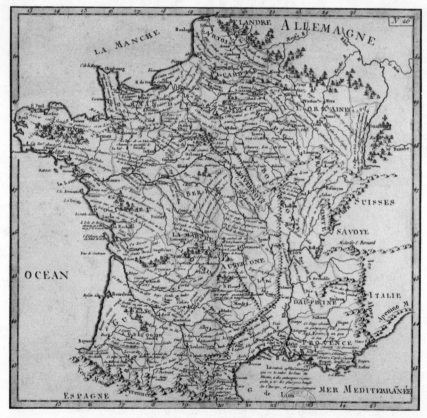

Figure 24 Commercial France. There is no known author or date for this map, which did not represent an improvement over a written summary of economic conditions. "La France commerçante." Phot. Bibl. Nat., Paris, Ge. D. 6818.

at a glance ("au premier coup d'oeil"). Equally typical, the author of the subject content (Marie Le Masson Le Golft) displayed information on a background map made by a geographer, often for another, more general purpose (here by Maurille-Antoine Moithey). Le Masson Le Golft, a teacher in Le Havre, considered this map to have pedagogic value, encouraging the reader to conceptualize relationships between social, cultural, and anthropological patterns and environmental conditions. On the map, areas of the world inhabited by nonwhites were shaded, to distinguish them at a glance. Le Masson Le Golft differentiated by symbols subcategories within four thematic groups: religions, customs, skin color, and physical form (fig. 25). By matching symbols covering a land area with the legend, the reader can learn that the French are Catholic, are educated, live in communities, and are kind, well proportioned, and handsome. The symbol for being educated was a triangle; for living in communities, a bee; for savage behavior, an arrow; for practicing polygamy, four dots; for practicing nudity, a squiggly line; for being kind, a dagger pointing

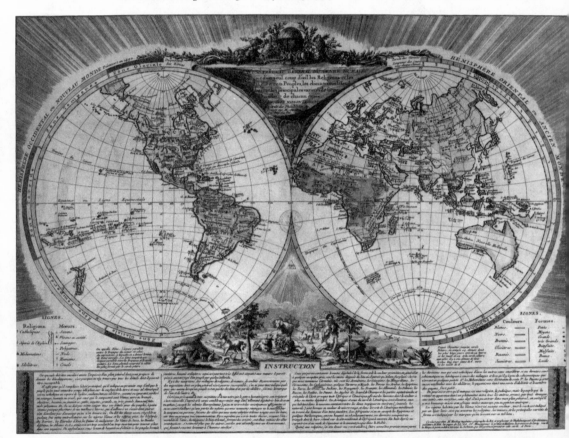

Figure 25 Ethnographic map of the world. The outline map of the world was by Maurille-Antoine Moithey; Marie Le Masson le Golft added information that would be classified today as related to cultural and physical anthropology. "Esquisse d'un tableau général du genre humain . . .," 1794. 1;55,000,000; 67.5 cm × 50.0 cm Phot. Bibl. Nat., Paris, Ge. C. 8674.

down. Nuances were given by including ten or more symbols for the same theme, so that we learn that both Christians and idolators live in China.

An incomplete economic thematic map from the late eighteenth century has survived in manuscript form.[5] The map (which may have been derived in part from works by Robert de Vaugondy and Robert de Hesseln) listed about one thousand localities (fig. 26). All cities were given symbols suggesting walls, a palace, and a cathedral. Each was surmounted by as many as fifteen conventional signs indicating ecclesiastical or civil institutions. The map's chief innovation was the use of symbols for economic resources and products. Agricultural symbols were provided for wheat, other cereals, vineyards, linen, pastureland, woodland, heath, cider, beer, sheep, horses, poultry, and so on. The productive capacity of a province was shown by the multiple of a given symbol under its name. Battlegrounds, fortified places, great châteaus, and brief phrases identifying such local curiosities as the tomb of three murdered children in Langres or the relaxing charms of Belle-Ile also appeared. Given the importance of domestic commerce in debates over taxation, public works, and the operation of markets, it seems odd that transportation routes were omitted (rivers and bridges alone appeared). Perhaps the mapper simply lacked room. One of the great achievements in French thematic cartography in the nineteenth century was the creation of a means whereby centers of production, transportation routes, and the volume and direction of commodity traffic could be represented in simple graphic terms. Before that could be accomplished, a way had to be found to convey information more efficiently.

Intellectual, social, and economic changes may well have encouraged experiments with thematic maps, seeking to better understand and analyze contemporary developments, but the creation of thematic mapping was by no means the inevitable product of converging circumstances.[6] We have already seen how the contour line, once invented, was for all practical purposes forgotten for a generation, only to be reinvented by Meusnier and Du Carla. When thematic mapping began to develop about 1800, the isometric line was already back in use, but it was not reconceived for thematic maps until forty years later. This example suggests that the search for appropriate techniques for thematic cartography was at least as much a matter of ideas as of technical skill. The general map, whether a geodetic or a topographical survey, treated space in static terms. Such a map might have to be changed as new roads were constructed, but modifications of this kind to the landscape occurred slowly and did not invalidate most of the features represented on a map. If anything, methods of making such maps changed more in a half century than the spaces represented on them. Thematic maps emphasized change by focusing upon economic activities, social behavior, and environmental conditions that fluctuated across space and over time. Thematic maps encouraged comparisons between similar circumstances in different places at the same time and at intervals, as more evidence was gathered.[7] National map surveys were made from intensive, costly, and lengthy field trips that were repeated perhaps once every generation. Thematic maps were developed from statistical records compiled by administrators, engineers, scientists, and

Figure 26 Economic map of France. A variety of symbols and brief texts describe economic conditions. The identity of the author of this undated map is unknown. Parts of the Austrian Netherlands have been filled in, but southeastern France was left incomplete. Approx. 1:650,000; 133 cm × 133 cm. Phot. Bibl. Nat., Paris, Rés. Ge. A. 1106.

persons in the liberal professions, often in the course of their regular activities, at little additional expense of time and money.

Had William Playfair (1759–1823) received greater recognition for his innovations in statistics, thematic cartography in France and elsewhere might have developed more rapidly. As a banker, journalist, publisher, real estate promoter, and statistician,

Playfair tried many ways to gain fame and fortune. He failed, but not for want of effort. Perhaps his lack of formal schooling hurt his reputation in the circles he wished to move in. His brother John, a famous mathematician, taught him that subject; he gained practical knowledge working with Andrew Meikle (inventor of the threshing machine) and James Watt. William Playfair acknowledged that early in life his brother John had made him "keep a register of a thermometer with the variations expressed by lines on a divided scale and also taught him that whatever could be expressed in numbers might be represented by lines."[8] From Watt he would have become familiar with the first automatic registering device, which drew a line showing the relation of steam pressure in a cylinder to the movement of the piston. Playfair invented a graphic method he called "lineal arithmetic," examples of which were published in his *Commercial and Political Atlas* (1786). The concept embraced broken-line graphs, bar graphs, circle graphs, and pie diagrams (the last three being Playfair's invention). His pie diagrams of 1801 were circles proportional in size to the areas of several European countries, colored green if the nation was a sea power and red if a land power. A red line to the right of each circle gave population in millions against a vertical scale; a yellow line to the right gave revenue in millions of pounds sterling. Playfair gave the number of inhabitants per square mile as well. Many of his graphs do not look tentative or experimental, partly because he was a skilled draftsman, but mostly because he understood the analytical properties of these graphic devices. He applied his methods to information about nations, the largest social units, at a time when most other graphic aids were limited to discrete natural phenomena such as the height of a river or the daily fluctuation of temperature. Playfair himself was aware that his interests were unique.[9]

Although Playfair made graphs, not maps, his graphs were based upon comparisons between geographical units, and he called one of his publications an atlas. Playfair believed that the geographic outline of a country was an unnecessary distraction in a graphic display of statistics categorized by nation, so he preferred to substitute a line on a graph or a shape such as a circle that could be varied in size to show wealth, area, population, or some other quantitative variable. Another pioneer in the design of statistical maps was A. F. W. Crome. Crome, a teacher of geography, history, and statistics at Dessau and Giessen, made a map dated 1782 using symbols to show the distribution of products across Europe. In 1785 he made a chart that compared the sizes and populations of all the European states. Other displays used superimposed squares to compare areas and represented population density with proportional circles scaled to units of area per person—the smaller the circle the greater the density.[10] Crome and Playfair effectively converted geographic units into other graphic forms but only Playfair's work was disseminated at that time in France, where it had been translated. Playfair claimed he was received with enthusiasm at the Paris Academy in 1789. Probably he was better known at that time in France than in England; no English statistician or economist called attention to Playfair's work until 1879.[11]

Even though Playfair did not experiment with thematic maps per se, his work suggested a direction that others might follow. The advantage of "lineal arithmetic," wrote Playfair, "is not that of giving a more accurate statement than by figures, but it is to give a more simple and permanent idea of the gradual progress and comparative amounts."[12] Playfair understood a fundamental conceptual equivalency to exist between graphs and maps: "The advantages proposed by this mode of representing matters are the same that maps and plans have over descriptions and dimensions written in figures; . . . for, whatever quantities can be expressed in numbers may be represented by lines; and where proportional progression is the business, what the eye does in an instant, would otherwise require much time."[13] Yet Playfair knew that graphs (or, let it be added, thematic maps) might defy comprehension at first. He wrote, "the reader will find, five minutes attention to the principle on which they are constructed, a saving of much labor and time; but without that trifling attention, he may as well look at a blank sheet of paper."[14]

During the period from 1789 to 1815 in France, many impressive surveys of demographic, economic, and social conditions were initiated.[15] A debate about the value of Playfair's work broke out among the many individuals involved in compiling and analyzing these surveys. The French distinguished between an English school of statistics reaching back to William Petty and John Graunt that emphasized quantitive work and a German school characterized by a literary method that produced judgments about economic and political conditions inductively drawn out of minutely detailed descriptions.

The Anglophile position was articulated by Denis-François Donnant, who translated several English books into French, including Playfair's *Statistical Breviary* (1801; trans. 1802). Playfair, in turn, translated Donnant's *Statistical Account of the United States of America* into English. Donnant's original and important contribution to the field of statistics was his *Théorie élémentaire de la statistique* (1805). In it he identified three categories of statistics: statistics that compare countries; statistics that reveal the resources, topographical features, and social (the nineteenth-century term was "moral") development of a single country; and statistics that differentiate among departments, districts, counties, or provinces. Donnant believed that graphic aids were critical to understanding statistical information because they synthesized data in a form that was easy to remember. Donnant also founded the first Statistical Society of Paris in 1803; its forty-two members were divided into six sections: medical and physical topography; weather and natural history; population and charitable organizations; rural economy; public works, industry, and commerce; and education and the fine arts.

J. Peuchet and P. E Herbin attacked Donnant and Playfair.[16] In their eyes Donnant was guilty of popularizing and rendering intelligible information about the power of states that should be restricted to government circles. A sound knowledge of men, they thought, provided leaders with a surer basis for public policy than statistics or a familiarity with the exact sciences. Peuchet and Herbin wanted to separate statistics from geography, restricting the former to highly variable institutional, social, and economic conditions and the latter to permanent natural and physical phenomena. Quantitative precision, they pointed out, cannot give science the authority of moral law. Donnant thought that statistics would discourage statesmen from engaging in wars by informing them about economic and demographic limitations on power; Herbin and Peuchet thought that statistics encouraged people to have opinions on matters they were not qualified to judge.

The disagreement was not resolved on its merits. Instead, a bureaucratic struggle for influence and authority, tensions between central and local government, and the succession of regimes determined whether individual officials were in a position to act on one point of view or the other. Under such conditions, projects that might have led to the production of thematic maps were abandoned at an early stage. Surely this situation limited the impact of Playfair's work in France. There is only one known French example from this period of a graphic aid modeled on Playfair's work: a civil engineer named Pierre- François Frissard made a graph of the value of the assignats and of revenues from 1789 to 1807.[17]

Yet two remarkable thematic maps were made in early nineteenth-century France. Neither has ever appeared in a scholarly or popular study, and so they have remained virtually unknown since the time of their making. These maps did not influence other mappers or map users then or since. The first was made by J. L. Wolff.[18] Nothing is known about Wolff or his map except the map itself. It was entitled "Essais de carte géologique et synoptique du Département de l'Ourte et de ses environs" and was made in 1801 (plate 5). It contained a great variety of symbols, lines, and colors. A red line with small crosses marked the departmental boundary, a gray line with dots and dashes marked the boundary between French- and German-speaking regions, and other colored lines separated geological regions. Mineral sources, mines, and the like were shown by various symbols. One dashed line indicated portions of waterways that could be improved for navigation, one dotted line showed the location of the ancient Roman highway. Numbers scattered on the map indicated altitude above sea level or population size. Perhaps the most remarkable symbols separated cities into three categories: manufacturing or trading centers, growing cities (ville florissante) and declining cities (ville déchue). Wolff's map related social and economic conditions to political and geological boundaries and showed an appreciation for urban trends, but it did not represent quantitative information precisely. Nevertheless, it should be appreciated for its astonishingly complex synthesis of different kinds of information and for its imaginative use of symbols, lines, and colors. To my knowledge, it was more sophisticated than any thematic map yet made.

Wolff's apparent interest in correlating geological information with political, social, and economic variables was in the air of the times. Jean-Baptiste Omalius d'Halloy (1783–1875) and Charles-Etienne Coquebert de Montbret (1755–1831) were preoccupied by a similar concern when they collaborated on a thematic agricultural map. As if to compensate for the paucity of information about the Wolff map, the archives are unusually rich in materials related to the Coquebert-Omalius enterprise. This is fortunate because the map itself in manuscript form, now in the Bibliothèque Nationale, does not provide a date, an explanation, or an account of the methods by which it was made (plate 6).[19] The map is a printed image of France, the original title of which has been covered with a new, handwritten title identifying it as a "minéralogico-agricole" map of France. Coquebert de Montbret's name appears, together with a key identifying colors corresponding to three hand-colored lines on the map that mark the northernmost limit of cultivation of orange trees, olive trees, and vineyards. A search in the archives of the Academy yielded the text of a speech Coquebert de Montbret gave on 19 February 1821 when he presented the map to its members. In the speech he acknowledged the assistance of Omalius d'Halloy, and the latter's presentation to the Academy a year later and his publications brought to light additional information on their collaboration. Finally, the map was the subject of a series of letters between the two men, now in the archives of the Académie Royale des Sciences in Brussels.[20]

Baron Charles-Etienne Coquebert de Montbret founded the *Journal des Mines*, taught physical geography at the Ecole des Mines, and served as a diplomat. His administrative and intellectual background prepared him to serve as a member of the Council on Weights and Measures, as director of customs and tolls on the Rhine, as director of statistics at the Ministry of the Interior from 1806 to 1808, and as secretary general of the Ministry of Commerce from 1802 until its dissolution in 1814. At the last post his duties were "the registration and distribution of despatches, the oversight of matters reserved by the minister for his personal action, and the charge of the ministerial archives."[21] He organized the most detailed surveys on industry and agriculture in 1806, on plant, animal, and mineral products in 1811, and on agriculture in 1814.

In his speech to the academicians, Coquebert de Montbret stated that the information depicted on the map for the northern limits of cultivation had been collected by the Ministry of the Interior in 1808 and 1809. At about that time, Coquebert de Montbret read an essay by Omalius d'Halloy describing geological patterns between the Pas-de-Calais and the Rhine. He then invited Omalius d'Halloy to make a map of the soil types and geological conditions of France. Coquebert de Montbret rejected administrative or political divisions in statistical analysis as too impermanent; he also rejected the more popular division of physical geography into watersheds, preferring the more complex and less familiar one of geological, subsurface patterns. It was his idea to superimpose information about crops onto a geological map as a means of determining the influence of soil conditions on agriculture. He believed that many conditions attributed to the influence of climate were in fact due to geological and soil patterns. Together, Omalius d'Halloy and Coquebert de Montbret conceived of a comprehensive study of commerce, physical geography, and statistics for much of Europe, but the collapse of the Empire shattered their positivistic, progressive assumptions—not to mention the information-gathering bureaucracy—upon which their effort was grounded. After 1815 Coquebert de Montbret was out of power, and Omalius d'Halloy became governor of the province of Namur in the enlarged kingdom of the Netherlands.

By the time he left office, Coquebert de Montbret had marked the northern limits of cultivation by commune on special maps of the departments; he also had a list of communes at or near these limits and a reckoning of the area under cultivation and the size of the average crop. (In his remarks in 1821, he said he intended to deposit these documents at the Institut de France, but they cannot be found in its library today.) Meanwhile Omalius d'Halloy, using much information forwarded by Coquebert de Montbret, had completed a geological map of France that also covered part of Belgium and Germany. In 1820 or 1821 he had it engraved and published.[22] Coquebert de Montbret then reduced the information on agriculture gathered in 1808-9 and transferred it onto Omalius d'Halloy's map.

By bringing his map to the attention of the Academy in 1821, Coquebert de Montbret hoped to arouse interest in making a second map. A comparison between two maps based on surveys a dozen years apart, he argued, would permit analysis of any changes in the northern limits of cultivation. He understood that such a comparison would indicate whether the climate was getting warmer or colder and would promote a better understanding of the relative impact of meteorological conditions and geological patterns on agriculture. He clearly grasped the analytical power of thematic maps to illuminate changing and permanent conditions and to distinguish between causal and coincidental relationships. The Academy asked that two copies of his map be made, but it is impossible to determine whether the map in the Bibliothèque Nationale is the original or one of the copies.

Their correspondence preserved in Brussels shows that at the same time, Coquebert de Montbret and Omalius d'Halloy wanted to expand and publish the latter's geological map as a French geological survey, equivalent to geological maps of England recently published by William Smith (1815) and George Bellas Greenough (1819) or of the Paris region by Georges Cuvier and Alexandre Brongniart (1810). (Pierre Dufrenoy and Jean Elie de Beaumont did not begin their geological survey of France, more detailed than the one Coquebert de Montbret and Omalius d'Halloy proposed, until 1825.) The map Omalius d'Halloy published in 1820 or 1821 was too small to contain all the information that had been collected; Coquebert de Montbret had recently made trips to Perpignan and Marseille, to Brittany, and to the Cévennes, Grenoble, and Briançon, during which he had more detailed observations. Coquebert de Montbret suggested they contract with a map printer who had an accurate base map of France on which geological and soil types could be superimposed and colored. He proposed Charles-François Delamarche, who then possessed the plates of Pierre de Belleyme's map of France. Apparently Delamarche was willing to publish and sell their map and offered to give the two authors twenty or twenty-five copies each. Omalius d'Halloy came to Paris in the late summer of 1822, at least in part because he and Coquebert de Montbret wanted to conclude their map-publishing discussions. The project fell through after Omalius d'Halloy returned north. On 3 November 1822, Belleyme's maps were sold at auction.

Nothing remotely similar to Coquebert de Montbret's map had been made before. In 1823 J. F. Schouw, a Danish botanist, published a book on the fundamentals of plant geography that was illustrated by twelve sets of paired Eastern and Western hemispheres on facing pages portraying by colors the vegetative cover of the earth.[23] Schouw's maps may well have been the first of their kind to represent global distributions, but they were clearly preceded by Coquebert de Montbret's map of France. Robinson's statement that Schouw's work was the "first real attempt to map the distribution of vegetation" must now be qualified.[24] Moreover, Coquebert de Montbret's map was based on a comprehensive survey, whereas Schouw's obviously was based upon the analysis of many written sources.

By the 1820s, the impetus to make thematic maps based on the *physical* world had ebbed in France. The reception the academicians gave the work of Coquebert de Montbret and Omalius d'Halloy was cordial but perfunctory. The Academy did not perceive the development of thematic mapping as an opportunity for itself and its members, as it had seen the development of geodesy under the Cassinis. The great geodetic survey of France had been associated in the minds of scientists with an understanding of the shape and size of the earth and with the search for an accurate means of determining longitude, especially at sea. The development of the chronometer and the introduction of metric units of measure through geodetic fieldwork appeared to satisfy their curiosity. When in 1806 François Arago and Jean-Baptiste Biot began their verification and extension of the measurement of the meridian initiated by Méchain and Delambre in the 1790s, an end to an era was reached.

Alexander von Humboldt gained great fame for his efforts to gather information revealing regular, systematic, intelligible patterns in the natural world of physical and biological geography. His belief in the unity of nature and his scientific expeditions excited the French, with whom he worked for some twenty years at the very beginning of the nineteenth century.[25] Humboldt inspired Arago and others to record electromagnetic observations with consistent instruments; their program, supervised by the Bureau des Longitudes from 1810 to 1817, was the longest sustained scientific research program at that time.[26] Some of the data were later plotted and mapped as isodynamic lines, according to Humboldt's model. (In 1804 Humboldt made a maplike diagram showing the varying intensity of a magnetic field, and in 1816 he made a similar image of isothermal lines of average annual temperatures on which the prefix "iso- " was first coupled with a descriptive term).[27] Notwithstanding the immediate, positive impact of his activities and ideas on French science, Humboldt never attracted disciples or created a school in France. After Humboldt returned to Germany in 1827 and Arago entered politics in 1830, no one in the natural sciences picked up where they left off. In France, at least, scientists increasingly preferred to study other topics that happened to lack a cartographic dimension.

One of the few original French contributions to thematic cartography involving physical geography used colors to express rank order or the significance of particular features. Entitled a "carte typo-graphique," it was lithographed in 1823 by the printer Firmin-Didot, perhaps as a model to interest editors of atlases. Blue showed the sea, major rivers, mountains over 5,500 meters above sea level, and major highways; red showed canals, rivers that are tributaries of major rivers, mountains between 4,500 and 5,500 meters in elevation, and some second-class roads; brown, dark green, and pale green showed rivers, mountains, and roads of decreasing importance or size. Foreign countries, departmental boundaries, and military districts were shaded in yellow; prefectures and subprefectures, cities, and the limits of navigability on rivers were displayed in black.

In the 1820s and after, other Europeans were more interested than the French in generating information about the physical world and synthesizing knowledge of physical geography in maps and atlases. Publishing ventures worthy of the heritage of the great French map printers of the eighteenth century were launched in Belgium by Philippe Vandermaelen, in Germany by Heinrich Berghaus and August Petermann, and in England by Alexander Johnston. Two Danes won a significant competition in 1825 sponsored by the Société de Géographie de Paris (founded in 1821). The society's first competition to expand geographical knowledge was on mountains. Although no map was called for in its announcement, which referred only to mountain chains and elevations, the second entry was accompanied by a manuscript map submitted by two Danes, O. N. Olsen and J. H. Bredsdorff, who were awarded a six- hundred-franc gold medal. They then compiled a more detailed map, which was published in 1833.[28] Olsen did most of the work on the second map. It was the first topographical contour map to cover Europe. Two colored versions of the first state of the copperplate engraving were made, one emphasizing five rock types, the other hydrographic basins and divides; on the second state, colors were used to differentiate mountain systems. Olsen was a military officer, surveyor, and instructor in surveying and drawing who eventually directed the topographical section of the Danish General Staff. He had probably become familiar with the technique of contour lines on an earlier visit to France. This example highlights the enormous progress people in other countries soon made. The growth of specialized scientific societies in all European countries, the establishment of professional mapping offices in government, and the international diffusion of French books and educational approaches to engineering enlarged the number of institutions and individuals able to produce maps of high quality and originality.

Thematic cartography in France after 1820 emphasized social and economic phenomena. Engineers, not scientists, made the most significant contributions to thematic cartography. The role of engineers was a legacy of engineering education and practice of the late eighteenth century. Some French engineers did their most important work in thematic cartography in the 1840s and 1850s, two decades or more after they had been trained in schools that included cartography in their curricula. Ironically, by that time the role of cartography in engineering education in France had already been deliberately circumscribed. Before pursuing the development of thematic cartography in France after 1820, let us first consider the role of cartography in the education of engineers in France and the reasons it changed.

Cartography and Engineering Education, 1747 to circa 1830

The curriculum of the Ecole Royale des Ponts et Chaussées established a role for cartography in the training of engineers. The school began in 1747 in the drafting office ("bureau des dessinateurs") of the administration of the Ponts et Chaussées under the direction of Jean-Rodolphe Perronet. Initially this office supervised and coordinated the work of engineers in the field by evaluating maps they made and sent to Paris and by recording progress in road and bridge building on maps made in Paris. At first Perronet simply wanted his students to learn how to make the kinds of maps they would have to provide in their professional work. In time he added tutorials in a variety of subjects. After 1763, Antoine de Chézy supervised the school's day-to-day operations with the help of two engineers, one of whom was responsible for engineering drawing and mapmaking. There was no other permanent faculty until 1804; one feature of the school was that the top students shared the teaching with part-time staff.

Pressures to modify and regularize the school's curriculum were channeled by Turgot in 1775 into a major reform effort.[29] (Turgot tried unsuccessfully at this time to eliminate the corvée, or use of forced labor in public works projects.) After 1775 courses were divided into one group of required studies and another of electives; summer and winter field experience, always a requirement, was limited to permit students to complete regular courses; Perronet's practice of using the best and most advanced students as tutors for the others, at first a practical necessity, was maintained for the pedagogic value of peer teaching; and grading, another older practice of Perronet's, was standardized, as were competitive entrance examinations and final examinations. Enrollment, which had fluctuated between 16 and 110, was limited to 60, divided into three classes of 20 each. In practice most students took longer than the optimum three years to finish. They studied geometry, trigonometry, differential and integral calculus, elementary and advanced surveying and mapmaking, mechanics, hydraulics, conic sections, theory of curvilinear surfaces, stereotomy, strength of materials, and architectural and structural design. In addition, they continued to make maps at various scales for use by the administration of the Corps des Ponts et Chaussées.[30] Students advanced if their projects were judged favorably by the assembly of the Ponts et Chaussées, its highest body, and by outside specialists. They earned credits toward graduation according to a schedule that awarded a certain number of points for work in a specific field by courses taken and by juried projects, with a sliding scale for the latter according to the student's grade. Cash prizes were also given to the two best students in each subject competition. The credits of all students were computed so that the best three in each class could be selected to serve as teachers. This meritocratic system produced most of the 230 commissioned engineers serving in the Ponts et Chaussées in 1785. The school's curriculum, based on a judicious blend of theoretical and practical studies, was widely recognized as providing the best training in civil engineering in the world. Perronet aspired to prepare engineers able not only to build a bridge, a road, or a canal, but also to design urban street patterns and civic monuments—in other words, professionals who combined mathematical analysis and artistic style. The Revolution shook this carefully constructed edifice, but like a well-designed bridge with a generous safety margin, it survived intact.

The preponderant weight given in the curriculum to architectural, mechanical, and mathematical studies gives a misleading impression that cartography was a minor and rather incidental subject. Cartography was conceptually integral to the design process. As Brooke Hindle has observed, many of the great engineering designers of the early industrial revolution (e.g., James Watt, Benjamin Latrobe, William Thornton) had considerable experience in mapmaking.[31] The same qualities of visual design that made a map intelligible and attractive were desirable in engineering drawing. The ability to draw well was more than a means for expressing an idea; it could also be critical to the very conceptualization of the idea. The processes of invention and design were essentially visual, whether a craftsman learned through emulation or an engineer with a technical education studied the interrelated features of a machine or structure at a desk. From the nonverbal, graphic elaboration of forms on paper, the engineer understood intuitively whether a certain element or structure fitted or was strong enough. Graphic design complemented empirical knowledge and calculations derived from theory. When Eugene S. Ferguson wrote that the "nonliterary and nonscientific" intellectual component of technology "has been generally unnoticed because its origins lie in art and not in science," he made the point that some aspects of technical design rest "largely on nonverbal thought and nonverbal reasoning," which is largely pictorial.[32] Training in cartography was essential because engineers needed to make maps in the field. It also encouraged them to think more critically and creatively about their work. The kind of spatial thinking they learned from cartography was easily transferred to other problems and situations.

As part of their education, students had to make maps of an area in the Paris region (the Montmartre hill was a frequent subject) and design a road for it. This kind of field survey closely approximated a situation they would encounter in professional service. The school's curriculum also reflected an appreciation of cartography for its heuristic as well as its practical value. As part of their final examination, students were asked to imagine a territory large enough to include forests, châteaus, cities, rivers, canals, roads, and farms and then make a map of it as if it actually existed. About thirty of these maps have survived.[33] They represented a kind of abstract, pure spatial thinking. In these maps, as in a game, the students created artificial situations in which elements of reality could be carefully manipulated. Through play, people can discover possibilities they might otherwise overlook. Such an activity might influence how the students would conduct themselves at work. In effect, they were being asked to consider how individual projects can, cumulatively, effect a spatial transformation on a large scale. The maps portray a commercially prosperous, agriculturally productive country in which institutional and physical obstacles to trade have been eliminated and rural and urban settlements coexist harmoniously. This map exercise taught the students that using their imaginations was a vital part of their work. The practical field survey for a single project and maps of imaginary territories were essentially complementary. In this way the students of the Ponts et Chaussées were taught to appreciate the relation between aesthetic and structural aspects of design (plate 7).[34]

Initially the Revolution had a greater impact on the organization of the Ecole Royale des Ponts et Chaussées than on its curriculum. Entry competitions in 1791 were to be conducted by department, but administrative difficulties at the school and in the provinces prevented their being held. Because no new students came to the school and because many of the students already there left to serve the state, enrollment fell.[35] When Perronet died in 1794 no one was appointed to the position of "premier ingénieur," the ex officio head of the school. Jacques-Elie Lamblardie, who was born the year the school was founded (1747) and graduated from it at the age of twenty, had distinguished himself in the field of hydraulics with proposals to deepen and keep clean the approaches to Dieppe, Le Havre, and Rouen. Lamblardie was appointed the school's director without other duties in the civil engineering corps on 28 February 1794. He soon became involved in the establishment of a new engineering school. The state needed more engineers; many army engineers (all aristocrats) had emigrated, and France needed engineers for defense and for public works. In September 1793 the merger of the army's Corps de Génie Militaire with the civilian Corps des Ponts et Chaussées had been proposed. This would have permitted the assignment of engineers to the places they were most needed. A law of 11 March 1794 considered establishing an Ecole Centrale des Travaux Publics by enlarging the existing Ecole des Ponts et Chaussées to include the training of both military and civil engineers. In September 1794 this proposal was revised to make the new school independent of any existing institution, and a year later it was given the name it bears today, the Ecole Polytechnique. Lamblardie then had to defend the autonomy of the civil engineering school as an advanced program for graduates of the Ecole Polytechnique.

The Ecole Polytechnique therefore was established as a basic institution whose graduates could then apply for admission to other schools for more specialized training.[36] This division of teaching assignments affected cartography, since all engineers were supposed to acquire knowledge of its fundamentals at the Ecole Polytechnique. But the role of cartography in the curriculum of the Ecole Polytechnique was far less important than it had been in the pre-1793 Ecole des Ponts et Chaussées. During the period between the foundation of the Ecole Polytechnique and the 1820s, the theory of building science was transformed by several French engineers. The new school embraced Gaspard Monge's descriptive geometry, which previously had been taught only at the army's engineering school. Descriptive geometry was conceived as a method for portraying three-dimensional bodies on a two-dimensional surface, rendering the picture approach superfluous by eliminating the need for perspective. It emphasized analysis of form through geometry in a way that could be applied to a wide variety of design problems. Jean Rondelet, Jean-Victor Poncelet, Claude-Louis Navier, Jacques-Nicolas Durand, and Gauthey built upon Monge's method (and upon developments in mathematics and mechanics by Coulomb and others) to translate into practice more powerful theoretical models in civil engineering and architecture by making it possible to solve structural, organizational, and financial problems through drawings and calculations.[37] Engineers trained in these techniques did not need to know much about cartography to perform competently. The Ecole des Ponts et Chaussées before 1793 emphasized cartography as an exercise in spatial thinking; at the Ecole Polytechnique, cartography was subordinated to stereotomy, or descriptive geometry. Together with linear and aerial perspective, engineering drawing, and measurement techniques, cartography was taught as an ancillary subject without wider ramifications. Apparently visionary maps were no longer called for, either at the Ecole Polytechnique or at the Ecole des Ponts et Chaussées.

A survey of curricular documents reveals what happened to cartography in the Ecole Polytechnique. At first maps were regularly required for road projects and for civil and military architecture, as much to emphasize draftsmanship as for understanding the principles of cartography. The published outline of the course in topography in the first decade of the nineteenth century proceeded from the theory and use of repeating circles and other instruments in geodesy and elevation measurement to the study of cartographic projections and the theory of curved lines and solids, then to various techniques of making field surveys and representing landforms. Students were taught techniques adopted in 1802 by the Commission on Cartographic Symbols. Students had to make maps using hatch lines, isometric lines, and colored tints, which in turn were used in courses in military science for studies of fortification and battle tactics. Instruction was provided by specialists in military science and in descriptive geometry, assisted by tutors in topographical drawing. The course included fieldwork in the hills of Gentilly and Montmartre and the forest of Fontainebleau. A fourth map was added by 1808–9 for use in a course on the construction of public works, as the background for model roads and canals. Mapmaking was therefore not so much a subject in its own right as a topic related to other courses. This approach emphasized the utilitarian, applied nature of maps as a means of graphic communication.

As Napoleon's need for officers and engineers increased, so did his demands that the students of the Ecole Polytechnique learn cartography. Although students enjoyed drawing classes as a form of relaxation from more rigorous studies, the art of drawing was defended because it developed the hand, eye, and taste. To make room for more serious work in topographical drawing, the curriculum of the Ecole Polytechnique was revised to drop studies of mining and naval architecture and to reduce studies of civil and military engineering, since these subjects were taught more intensively in the applied schools.

At the army's school of artillery and fortifications at Metz, in 1803 a committee planned a curriculum to provide more intensive work in cartography to graduates of the Ecole Polytechnique.[38] The coursework covered topics in topography, geodesy, and map drawing already introduced at the Ecole Polytechnique, but in greater detail. Students also learned how to prepare descriptive reports and statistical surveys as these related to mapping surveys. They spent sixty days of fieldwork near Metz, working on a map for a highway. Each student was assigned a road segment eight-hundred meters long. They had to measure water level at fifty-meter intervals and land elevation at one-meter intervals and make maps with isarithms at a scale of 1:2,000. They spent an additional fifteen days on a geodetic survey of the environs of Metz, each student being given a certain number of places to tie in to a triangulation network. Twenty days were spent in more rapid, less detailed mapping exercises of an area four kilometers square. Finally, fifteen days were devoted to reconnaissance mapping in an area sixteen kilometers square. For that exercise, each student was given a basic geodetic map, to which he was supposed to add highways, footpaths, bridges, rivers, forests, and buildings.

Another of the specialized schools taking graduates of the Ecole Polytechnique, the successor to Prony's Ecole Nationale de Géodésie Pratique, was the Ecole des Ingénieurs Géographes, founded in 1795. It has been overlooked by historians both of cartography and of education; Frederick Artz, in his detailed history of technical education in France, described it in a single page. The school's three teachers (applied mathematics, mechanical and engineering drawing, landscape drawing) provided a two-year course of study attended by about twenty graduates of the Ecole Poly-technique and ten others.[39] A copy of the program from the year VIII has survived. The mathematical part contained trigonometry problems not covered in classes at the Ecole Polytechnique, as well as spherical trigonometry related to geodesy. It dealt with astronomy and with different ways of measuring the pendulum and the me-ridian, of determining latitude and longitude at sea and on land, and of orienting a map. Students also learned the theory of geodesy and of map projections and practical ways of constructing globes and of enlarging or reducing maps to different scales. Studies in the physical sciences embraced mineralogy (including the use of symbols and colors on geological survey maps) and aspects of physics relevant to instruments (the construction of lenses, the use of the barometer, etc.). The art of drawing included color conventions to indicate land use, lettering on maps, perspective for buildings, standardized representations of topographical features, and methods useful in making military reconnaissance maps and cross sections. Practical exercises with the repeating circle and other more common instruments were provided. Students were expected to use logarithmic and astronomical tables quickly and without error, to measure area from a map, and to compose maps from written and mathematical observations in the field. This school prepared people to organize, execute, and verify the accuracy of topographical and geodetic surveys at various scales and under many conditions, using several techniques and instruments. "The men who finished the course," wrote Artz, "took positions in the army or the navy or with the department of public works. Some became teachers of geography and map making in the higher military and naval schools. . . . It seems to have been difficult to get students to enter the school; the army and navy usually used members of their own staffs, who had learned their map making on the job, to do their map work for them."[40] The school was closed in 1802 but reopened the next year; its budget then was fifteen thousand francs. Between 1809 and 1813, when recruitment ceased, thirty-eight graduates of the Ecole Polytechnique entered it. Poor prospects for promotion in this field no doubt encouraged engineers to consider other areas. The council of the Ecole Polytechnique deplored this state of affairs in the year VIII. Nevertheless, some of the school's graduates had distinguished careers: Louis-Clair Saint-Aulaire became an ambassador and peer, and Edme-François Jomard shared the honor of creating the field of the history of cartography with the Viscount of Santarem.

Cartography did not retain its importance in the curriculum of the Ecole Polytechnique for long after the fall of the Empire. A latent tension existed between geodesy, which was related to astronomy and mathematics, and topography, which was related to the design of civilian or military engineering works. This meant that cartography existed as a hybrid subject exposed to different and potentially conflicting views of what to teach. In 1816 the quality of instruction in topographical drawing was already declining. By 1819–20 the school recognized that efforts to reverse this decline were unsuccessful. All the specialized schools receiving graduates of the Ecole Polytechnique were complaining about their lack of proficiency. This became a matter of concern at the Ecole Polytechnique at a time when the allocation of responsibility for the cadaster and for the national map survey was unsettled and when specialists were divided over appropriate topographical mapping techniques. According to a committee examining the issue, the condition of topographical drawing in the school was typical of education there in the graphic arts in general.[41] The committee considered raising admissions requirements for drawing, increasing the number of hours of study, and tightening other school requirements, only to conclude that these measures would be ineffective. Unable to impose uniformity in topographical drawing on all public services, yet acutely aware of differences in practice, the committee was uncertain whether the curriculum it adopted was appropriate and valid. It chose to wait until technical standards in topographical drawing involving hatch lines, contour lines, coloring and shading, and the like had been resolved in the public services. Some faculty members dissented from this decision out of fear— justified, as it turned out—that this approach would only delay curricular reform. A majority, however, agreed with the committee's decision. In effect, the Ecole Polytechnique failed to support cartography in engineering education at a period when its place in the curriculum depended primarily on the value the school's faculty placed on that subject. In 1820 the instructor in geodesy reported that students did not pay attention and were behind in their work. In 1827 a school committee recommended dropping fieldwork in geodesy because it was too lengthy, involved too many assignments, and did not interest the students. In 1831 required work in drawing was limited to buildings and machines; topographical drawing became optional.

The Ecole des Ponts et Chaussées could not compensate for the erosion of support for cartography in the Ecole Polytechnique. During the First Empire, the army took 75 percent of the graduates of the Ecole Polytechnique and the Ponts et Chaussées took 9 percent. During the Restoration, priorities shifted slightly; 21 percent went into civil engineering. Graduates of the Ecole Polytechnique accounted for about half the student body of the Ecole des Ponts et Chaussées; the other students were admitted after passing examinations in drawing, physics, and advanced mathematics.[42] After Lamblardie's death in 1797, Antoine de Chézy, Perronet's associate who had retired in 1790, took over the school. When he died in 1799 Prony was appointed to replace him. Prony died in 1838. For nearly forty years, therefore, the Ecole des Ponts et Chaussées enjoyed a continuity of leadership. Perronet and Prony, who together supervised the school for over eighty of its first hundred years, were strongly identified with cartography. But mapmaking at the school in Prony's years was not what it had been when Perronet was in charge. Prony refined the curriculum, making its subunits clearly related to routine aspects of professional engineering work. Thanks in part to Prony's own work as director of the cadaster in the 1790s and to administrative changes that followed under Napoleon and during the Restoration, map surveys for fiscal and geodetic purposes were established as permanent state activities, but in units in which civil engineers did not serve. This meant that civil engineers in the nineteenth century were less responsible than their predecessors for providing maps. They were also more ably assisted by larger numbers of better-trained subordinates familiar with surveying. Meanwhile, engineers needed to learn more mathematics and mechanical analysis in order to keep up with advances in science and with the development of new materials and machines. In the 1820s and after, cartography and engineering drawing were minimized in the Ecole des Ponts et Chaussées; there and in the Ecole Polytechnique, engineers trained in France in the 1820s were given far less exposure to cartography than any previous group of students.

The changing role of cartography in engineering education can be seen in the work of two engineers of the Ponts et Chaussées who made impressive contributions to thematic mapping, Léon-Louis Lalanne and Charles-Joseph Minard. They had been trained as engineers at a time when cartography was an important part of professional training; they made important contributions to cartography at a time when that topic had been superseded by mathematical and theoretical topics. Their principal contribution to cartography lay in the synthesis they achieved between spatial concepts and mathematical, analytical rigor in the design of thematic maps.

"Parler aux Yeux": Economic and Social Thematic Maps, 1820–50

The development of thematic cartography in France after 1820 responded to the twin processes of industrialization and urbanization. In the late eighteenth century topographical configurations such as watersheds figured prominently in the concepts of map-related studies of transportation routes, but heavier than expected rates of urban growth, rising construction costs, and competition between sponsors of major public works projects brought about a change in the views of some engineers. The realization grew that investment in public works should be determined by existing and probable concentrations of people and business rather than by purely topographical considerations, since only then could the investment be successfully amortized.

Those concerned with the consequences of urban growth and with investment policy included persons in the liberal professions as well as engineers. Better information about existing conditions and more sophisticated methods of analysis were appreciated as never before. The problems they studied with (and without) the use of maps were not unique to France. Individuals in other Western countries were at work on various kinds of thematic maps related to transportation, public health, and the weather. These maps were based upon increasingly rigorous and compre-hensive gathering of statistics, usually for some purpose other than making a thematic map. In some aspects of thematic cartography, Frenchmen did conventional work. This in itself is important, because the diffusion of information- gathering techniques and the production of thematic maps that were intelligible across cultural borders quickened the international exchange of information and emphasized statistical com-parisons between nations. But in some aspects of thematic cartography, Frenchmen distinguished themselves and created concepts and techniques that significantly ad-vanced the state of the art. For these reasons, the contributions of the French to thematic cartography in the period 1820–50 should be presented in a broader context.

In eighteenth-century France environmental conditions were associated with dis-ease. The source and spread of disease were major concerns of government, but conflicting interpretations of the pathology of contagion hampered prophylactic efforts involving environmental change. (Ironically, the belief that diseases such as yellow fever are spread by polluted air and water, though incorrect, led to needed environmental improvements.)[43] In 1786 the Société Royale de Médecine established a committee to conduct research on medical topography, and in 1833 the Académie Royale de Médecine established a permanent commission on topography and medical statistics. Between 1771 and 1830, over one thousand reports had been submitted on about nine hundred epidemics in 1,370 communes in seventy-two departments of France. Their content was transformed by locality into a table to bring to light information on soil and water conditions, personal hygiene of the population, dates for the onset and end of the disease outbreak in question, and the number dead, but a lack of uniformity in the reports impeded analysis. The record does not show whether anyone in France had yet conceived of a medical map for the purpose of recording and analyzing disease/environment relationships.[44]

The cholera epidemic of 1832 generated many medical maps in Europe.[45] Some cholera maps used dots to show where sick individuals resided. Others showed the spread of the disease across continents with colored lines and shaded areas keyed to dates. Many maps made at that time to illustrate books or reports have since been lost. Maps of epidemics of 1832 were made well into the 1850s, as successive outbreaks of disease spurred research that might reveal its pathology and suggest preventive measures. French maps of cholera epidemics have not appeared in the literature on medical maps, which emphasizes English materials. Two copies of a set of French maps have apparently disappeared, but mention of them at a meeting of the Aca-démie Royale de Médecine in 1833 suggests their contemporary importance. The Ministry of Commerce and Public Works had transmitted a map of France in twenty-four sheets that identified the localities where cholera broke out, together with the dates of outbreaks, to the Academy of Medicine and the Academy of Science. The medical academy's minutes do not contain a description of the cartographic symbols used but merely include some notes saying the maps were made by Ségur du Peyron, secretary of the Conseil Supérieur de Santé, who recorded information from daily reports of the departments.[46] Another French cholera map was made by a doctor who included it in a book he wrote about the epidemic of 1832 in Rouen. He used the spot technique, a dot on the map indicating a case and its locality. On one small street with twenty-five cases, the dots blended into a line. A table at the book's end was organized as a gazetteer and listed the number of cases by the street name.[47]

Finally mention must be made of a French medical map having nothing to do with cholera. In 1839 Joseph-François Malagaigne made a map of the incidence of hernias in France by shading departments according to a scale keyed to the national average. Malagaigne wanted to see whether a relation existed between the distribution of hernia cases and diet or topography. He therefore added two lines taken from an atlas by Conrad Malte-Brun showing the northern limits of olive and vine cultivation. He also differentiated between northern and meridional France along a line between La Rochelle and Geneva. Malagaigne found no correlation and had to reject his hypotheses, but he published his map anyway.[48] It survives as one of the oldest of a genre (shading by department) that varied little in style over the nineteenth century.[49]

The convergence of efforts in different countries is more apparent in the case of meteorological maps. Some Americans took the lead.[50] In 1840, in *The Philosophy of Storms*, James P. Espey published maps of storms from between 1780 and 1840 that noted wind direction and velocity. Elias Loomis made charts in the 1840s featuring temperature, pressure, wind force and direction, and precipitation.[51] Writing in 1843, Alexander Bache informed Lambert-Adolphe Quételet of "registers of the weather kept at a large number of our military posts and regularly reported to the surgeon general." He continued, "Prof Loomis" proposes a *meteorological crusade* and wishes . . . to obtain for a year hourly observations of the temperature and pressure of the air etc. all over the United States. What do you think of this! Would it not be as the Prof. says worth all the desultory observations we have been making for the past century?"[52] Joseph Henry, who taught science at Princeton University before becoming the first director of the Smithsonian Institution, was the first person to use the telegraph to collect meteorological observations, in 1849; the network linked five hundred stations by 1860. At the Smithsonian, charts were made to exhibit the distribution of temperature using isothermal lines in 1852, and synoptic charts of weather conditions were made daily as early as 1849.

Quételet no doubt endorsed the Americans' efforts. As the leading scientist of Belgium and a pioneer in the application of statistics to science and social issues, Quételet already had encouraged meteorological data-gathering operations.[53] Weather reports from observers in foreign countries had been published by the Académie Royale de Bruxelles in its *Bulletin* beginning in 1835. To master the data, Quételet made tables and *cartes figuratives*, a French term of the period for thematic maps. He used *isobariques*, or lines of equal pressure (reduced to sea level) to show barometric pressure in the summer of 1841 with data from twenty stations. Studying his maps, Quételet discerned three pressure zones across Europe and time/speed differences as weather systems moved across the continent. He also correlated mean and extreme temperature with the farthest limits of cultivation of vines and wheat.

Admiral Robert Fitzroy established an observation network in England in 1859. Fitzroy relied heavily on the telegraph to collect data, and he was the first to use it to transmit forecasts based on an evaluation of that data. The connection Fitzroy made between the weather patterns of today and of tomorrow, and between the weather in one place and in another, confused his contemporaries, who doubted that what he was doing was scientific.[54] The distinguished scientist Francis Galton made a series of ninety- three weather charts on his own initiative for the morning, afternoon, and evening of every day in December 1861, based on data from eighty stations.[55] Using abstract images composed of lines and dots to depict the extent of cloud cover and isarithms to depict barometric pressure, Galton showed how a variety of meteorological conditions could be related to each other analytically. He also experimented with a map of Europe divided into a grid of eighty-eight squares, each containing one of four colored symbols corresponding to a precisely defined range of barometric readings. He published his maps but noted that the printers had trouble meeting his standards. It would be interesting to know whether others encountered similar difficulties.

Perhaps the oldest surviving French weather map—and one of the oldest maps of an individual meteorological event made in Europe—was of the path of a storm across northern France on 13 July 1788. Its author and the sources of his information are not known.[56] Maps of the weather did not become common in France until the 1860s, as part of the activities of the national meteorological service. The service was organized by Urbain Le Verrier, a graduate of the Ecole Polytechnique who became an astronomer and director of the Paris Observatory. Information on wind, temperature, pressure, and cloud cover was noted at the Paris Observatory four times daily but was not displayed cartographically. In 1854 the minister of war asked Le Verrier to study a cyclone that had struck the fleets besieging Sebastopol. Le Verrier was able to reconstruct the cyclone's path. On 16 February 1855 Le Verrier presented Napoleon III with a proposal for a large meteorological network to alert mariners to storms. Two days later he presented the Academy of Sciences with the first maps of atmospheric conditions in France based upon information collected by telegraph. At first the officials of the telegraph authority were unwilling to establish a regular service for collecting data and distributing forecasts, but on 1 January 1858 Paris began to distribute a daily bulletin of weather conditions at fourteen French stations and five foreign stations. Le Verrier fulfilled his original plan to distribute warnings and forecasts in 1863. Then he began to make maps regularly. Most weather maps covered only Europe and omitted the Atlantic because data were transmitted only from land stations, but Le Verrier knew that most European weather systems came from the western ocean. To inform himself of Atlantic weather conditions, he consulted ships' logs. His maps of 1864 were published in 1866; weather maps were made daily in France after 1864.[57]

In meteorological and medical maps, the efforts of Frenchmen after 1830, though substantial, closely resembled mapping efforts in England and America. In social and economic maps, the work of the French was more original.

Charles Dupin (1784–1873) graduated from the Ecole Polytechnique, became inspector general of the Génie Maritime, wrote extensively about French and English commercial practices, was elected to the Chamber of Deputies in 1828, and was appointed senator in 1853. In 1819 he had become a part-time professor in the Conservatoire des Arts et Métiers. At that institution on 30 November 1826, Dupin gave an address on popular education and its relation to France's prosperity, during which he exhibited a map, "Carte figurative de l'instruction populaire de France," which he said appeared to have "initiated the cartographic portrayal of moral statistics," a term encompassing "a wide array of characteristics of populations."[58] Dupin's map was published (by lithography) in 1827 and remains the oldest known chloropleth map, that is, a map determined by civil or other nontopographical divisions "in which the enumerated data are assumed to be averages applying equally" throughout each district on the map, differences within each district being averaged (fig. 27).[59] Each district is shaded, colored, hatched, or dotted to indicate its statistical position relative to other districts. On Dupin's map the shading was not graded to a scale of tones with exact magnitudes corresponding to differences in the statistical values they represented.[60] Dupin was obliged to inscribe on the map figures for the number of persons in a department per male child at school. The darker the shading, the higher the number and thus the level of illiteracy. He concluded that more schools were needed in 4,441 communes in northern France and 9,688 in meridional France. Dupin's map showed a division of France into better-educated and more poorly educated parts along a line from Saint-Malo to Lyon.[61] (A Prussian manuscript map of 1828 gave a legend for shaded tones, but precise correlations between statistical variables and graphic techniques were standardized a generation later by Joseph Minard, of whom more later.) Dupin's contemporaries widely acknowledged the impact his map made on them.

One of the next social thematic maps received little attention, perhaps because its author, A. Frère de Montizon, was as unknown to his contemporaries as Dupin was famous. Frère de Montizon taught sciences and published educational items on history and philosophy.[62] His map deserved recognition as the first thematic dot map of population, yet his innovation had no effect on practice and was reinvented for a map dated 1859 by a Swedish army officer.[63] Frère de Montizon's map, dated 1830 and entitled "Carte philosophique figurant la population de la France," depicted France divided into departments with one dot per ten thousand inhabitants. A table in the margin identified departments by name (they were numbered on the map), population, and prefectural city. A few learned quotations were placed elsewhere. A straight line identified on the map as AB, drawn from the mouth of the Loire to Versailles and Charleville-Mézières, divided France into thermal zones, the north being where vines could not be cultivated. One of the author's preoccupations, as revealed by comments in the margins, was with the relation between population size, soil fertility, education levels, and criminal behavior, but his map, limited to population size, did not support an analysis of such correlation.

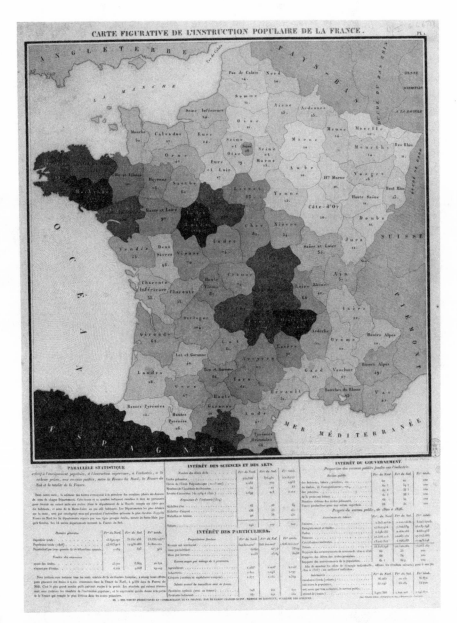

Figure 27   Popular education in France. This map shows that, in general, more people attended school in northeastern France than in the western and central regions. Darker tints on the map indicate a higher level of illiteracy. Charles Dupin's "Carte figurative de l'instruction populaire de la France" was printed by lithography in 1827 as an illustration in his *Forces productives et commerciales de la France*. Phot. Bibl. Nat., Paris, Ge. CC. 6588.

In 1829 Adriano Balbi and André-Michel Guerry made the first maps of crime that permitted the kind of multivariable analysis Frère de Montizon failed to achieve. Using criminal statistics for the years 1825–27 and the latest census, they made maps of crime against property, crimes against persons, and levels of education. Because the three maps were printed on the same page, together with tables of urban statistics on crime and education, the reader could easily determine for himself that France northeast of a line from Orléans to the Franche-Comté was better educated, that departments with high levels of crime against property had low incidence of attacks on people, and that among the departments with heavier levels of property crimes were many with higher levels of education. The partnership between Balbi and Guerry did not last, but each brought to it something the other lacked. Balbi was familiar with general maps, ethnography, and languages; Guerry was a lawyer interested in patterns of criminality.[64]

Lambert-Adolphe Quételet (1796–1874), the Belgian scientist and administrator who more than anyone professionalized the field of statistics and promoted its study and application, made many references to Dupin's work and was familiar with Guerry's. In 1829 Quételet did not include a map in his first published statistical study of population, society, and economy. In publications in 1831 and 1832, Quételet published three maps on the same themes as Balbi and Guerry had selected but covering a larger area—France, the Low Countries, and the lower Rhine. In 1833, Quételet received a letter from Guerry and a copy of the book he had written with Balbi. Guerry had written to Quételet as one of the few persons who might appreciate his work. In the margin of this letter Quételet reacted to the Balbi-Guerry maps, noting that high levels of education could not prevent crime. Quételet instead saw a correlation between well-developed transportation routes and high educational levels and emphasized ethnic and cultural variations to account for the statistical patterns of criminality.[65] In his maps Quételet achieved a more refined image than the French. Balbi and Guerry had shaded departments so that darker areas on the map had a higher incidence of a particular variable, such as crime or literacy. Quételet also used dots in proportion to statistical levels, but because his maps did not show departmental boundaries, changes in the tones of shading were smooth and continuous across the map's surface.

Quételet's maps were republished with the same technique in successive editions of his most famous work, *Sur l'homme et le développement de ses facultés, ou Essai de physique sociale* (first edition 1835). In this book Quételet argued that statistics are to the social philosopher what a thermometer is to a physician: if one knows the data for a healthy society, then statistical patterns will reveal disorders. Controversy surrounding Quételet's concept of the average man had an extraordinary and profound impact. Many people without a prior interest in thematic mapping probably saw a thematic map for the first time in an edition or translation of Quételet's book. Yet Quételet did not pursue thematic cartography after this initial, positive experience. Why? Perhaps he concluded that the analytic benefit of using maps as a background on which to project statistical information was not worth the effort. Perhaps, as George

Sarton has suggested, Quételet's genius had already fulfilled its highest potential by 1835.[66] In later years, Quételet devoted his energies to efforts to gathering statistics in many countries according to standardized methods. He was the creator of the International Congress of Statistics and presided over seven of its first nine meetings.[67] Thematic cartography was an important topic of discussion at most of these meetings.

Dupin, Balbi, and Guerry generated considerable interest in thematic mapping just a few years after Coquebert de Montbret and Omalius d'Halloy had received scant attention when they presented their agricultural map. The early thematic mappers, however, attempted nothing more than two or three maps, and such maps as they made were published separately or included in other books as supplementary material. The first major work organized around social thematic maps was Adolphe d'Angeville's *Essai sur la statistique de la population française considérée sous quelques-uns de ses rapports physiques et moraux*, published in 1836 in Bourg-en-Bresse. D'Angeville (1796–1856) made sixteen chloropleth maps of France divided into departments, covering population density, rate of population growth, number of farmers, development of industry, army rejections of draftees for poor health or being too short, life expectancy, adequacy of diet in grains, primary education, use of doors and windows in construction as an indicator of wealth, illegitimate births, foundlings, arrests for crimes, number of civil suits, number of persons failing to enroll for the draft, and incidence of failure to pay taxes. Who was d'Angeville and why did he make so many maps?

Adolphe d'Angeville's father was a progressive Jura landowner who encouraged improvements, and one of his sisters was the first woman to climb Mont Blanc. Adolphe spent ten years at sea as a naval officer, from 1811 to 1821. No doubt he became proficient in the use of modern sea charts. Upon returning home, he experimented with new crops and encouraged the development of rural producer cooperatives. Frequent trips to Paris in the 1830s gave him the opportunity to begin his book and collect statistical data from government sources.[68] Unlike Dupin, his political career in the Chamber of Deputies (1834–48) did not attract attention; neither did his book. Yet d'Angeville's book, which like Dupin's referred to the debate in France concerning social and economic development, was considerably more sophisticated than Dupin's in its use of thematic maps.

D'Angeville, like Dupin, saw France divided into modernizing and traditional zones, but d'Angeville understood better that the division changed geographically according to the statistical criterion of "modern" being used. And he included many more variables than anyone before him. His use of medical examinations from the army— to infer living conditions from the physical attributes of inductees—was truly original and has remained a theme of social and military studies ever since. His maps do not show a rigid partition of France but illustrate a series of trends.[69] Contrary to what Dupin and some others thought or suggested, d'Angeville believed that the division of France into a northern, more modern zone and a meridional, more traditional zone was far more a function of demographic and social factors than of the distribution of crops and of weather patterns. D'Angeville preferred the division of France according to river basins, not out of a belief in the transcending significance of topography such as men of the 1770s and 1780s held, but from an appreciation of their roles as corridors of commerce and culture.[70] His maps were less than satisfactory in two respects: they did not distinguish between social classes and groups (or between rural and urban communities) within departments; and each treated only one variable or criterion at a time. In text immediately preceding the maps, d'Angeville drew out some thematic comparisons between the maps. Criticisms of d'Angeville's book are perhaps unfair. He was limited by the quality of the statistical data at his disposal, yet he managed to combine statistical and cartographic analysis more successfully than anyone else, in France or elsewhere.

The book itself was divided into several parts. First each theme or subject was presented, together with an analysis of the statistical data and a set of conclusions. Then appeared two pages on each department, with statistical summaries for each subject analyzed. Next came tables arranged by subject and department, with notes on the sources of the data and on other relevant matters. The sixteen maps came at the end, preceded by a chart of cross-references directing the reader back to pages of text and tables relevant to each thematic map. The book was a synthesis of different kinds of materials and methods. Dupin and Quételet had used thematic maps conspicuously, but in isolated passages; d'Angeville conceived maps to be no more or less important than any other part. For him maps had the advantages of being unambiguous, of compensating for the dryness of numbers, and of calling attention to matters that would otherwise escape notice. Each map was shaded in five tones, one for every seventeen departments. More tones, more categories, d'Angeville believed, would have made the maps harder to understand. D'Angeville listed the departments in rank order in a table in the margin and indicated what the darkest tone on that map meant (e.g., lowest population density). The departments were numbered on the map in rank order as well. No place-names appeared on the maps; obviously d'Angeville assumed his readers knew that Bordeaux is in the Gironde, Lille in the Nord, and Lyon in the Rhône.

D'Angeville's book has achieved greater recognition today than it probably received when published. No one followed its example in France. Although individual Frenchmen developed social thematic mapping earlier than other Europeans, their activities did not become a self-sustaining movement. Clearly an institutional framework was not a precondition to creative work in this field, but its absence may have inhibited the development of thematic mapping once the work of these pioneers was at an end. In France no publisher or state agency provided the funds and continuity needed to produce economic and social thematic maps as rapidly as society changed. The creative French thematic mappers were highly self-motivated and received little support for their work. Perhaps the response of Quételet to thematic mapping is revealing of widespread opinions before the 1850s: after an initially positive and enthusiastic reception, the topic was dropped, no doubt because graphic design and printing techniques were not yet adequate to the conceptual demands and critical standards of statistical analysis. In this context, the work of two high-ranking engineers of the Ponts et Chaussées, Léon-Louis Lalanne and Charles-Joseph Minard, is especially significant. More than anyone else, they advanced the state of the art of thematic mapping, adding mathematical rigor and enhancing graphic design.

To appreciate what Lalanne and Minard achieved, it is useful to keep in mind comparable work of the same time by Henry Drury Harness in England and by Alphonse Belpaire in Belgium. Harness, a lieutenant in the Royal Engineers, made maps to illustrate a report of the Irish Railway Commissioners, published in 1837.[71] On one map he separated rural from urban populations, portraying the former by shading and the latter by proportional circles (their first known use on a map). He also published the first flow maps to illustrate the report. One showed the average number of passengers carried in one direction weekly by means of shaded lines, "the widths of which Harness clearly intended to be directly proportional to the number of passengers."[72] Harness's innovative maps apparently did not attract attention beyond the restricted circle of persons interested in the work of the Irish railway commissioners. Belpaire's flow maps can therefore be considered an independent innovation. He made two maps, published in the mid-1840s, of transportation movements in 1834, 1835, and 1844. He drew bands or flow lines scaled in units of transport, one unit being ten thousand tons of bulk merchandise, five thousand tons of baggage or packaged materials, or thirty thousand passengers.[73] He distinguished the nature of the traffic by coloring the flow lines differently. He used maps to make the point that railroads did not take traffic away from canals and that in the vicinity of towns there was heavy road carriage. Belpaire felt compelled to explain why he made maps. Like Humboldt and Playfair, he believed that mere numbers listed in tables did not promote comparative analysis and overall comprehension. Vision alone leads to an understanding of many factors as their interrelations are seen to form an overall pattern. Only someone interested in Belgian railways, however, was likely to take notice of Belpaire's work.

Belpaire and Harness did not pursue thematic mapping further. They lacked the opportunity to explore a range of subject matter and of techniques and to arrive at a more theoretical explanation of what they had done, so that others might understand their methods. Thematic maps by Harness and Belpaire call attention to two aspects of thematic mapping that greatly interested Lalanne and Minard: the representation of data where they actually occur, rather than by administrative units (such as departments in France), and mathematically determined scales for the color or shade and for the size of graphic representations.

Léon-Louis Chrétien Lalanne (1811–92) was the first person to make explicit the conceptual distinction between isometric lines and isopleths. Graphically, visually, they are indistinguishable from each other, but the differences are fundamental in thematic mapping. The only isometric lines in use in the 1840s showed elevation or underwater depths or magnetic and temperature observations, and only because very large numbers of observations had been accumulated scientifically. Humboldt, who understood the conceptual difference between a contour line of a hill or a bathymetric line of underwater depths and a curve line to show magnetic declination or temperature or atmospheric pressure, argued that such lines revealed patterns more clearly than tabular lists of observations.[74] The advantage of isometric lines over tabular lists in studies of nature, he wrote, was the same as the advantage of geodetic maps over lists giving the latitude and longitude of places. Humboldt was probably the first person to use the prefix "iso-" with a cartographic term; there are now over a hundred iso- terms in use.[75] Humboldt understood isotherms (lines of equal temperatures) to be a kind of isometric line, but he did not refine the concept further. Arthur Robinson has defined isometric lines as lines that "portray numerical distributions which may be termed basic, elemental facts. The values they represent actually can exist at points on the earth," such as elevation or temperature:

> whether the mapped data have been derived by single measurement, averaging of time series or some other statistical manipulation makes no difference. Isopleths and isometric lines look very much alike on a map, of course, but isopleths are different because they portray . . . relative values. These are higher order, more complex geographical concepts or abstractions that are a function of one element and space, such as density (e.g., persons per unit area), spacing, . . . geographical ratios. . . . [The] values on which the lines are based cannot actually exist at points.[76]

With Robinson's definitions in mind, we can follow Lalanne's intellectual steps.

Lalanne elaborated the distinction between isometric lines and isopleths in the course of his work at the Ecole des Ponts et Chaussées, which he directed after Prony. In the first of two reports to the Academy of Sciences, Lalanne approached the problem of determining many variables in an equation by representing numerical relations as straight or curved lines on a graph, so that by a single reading one or more unknown variables could be determined when the others were given. Lalanne understood the visual resemblance between his graphic representation of a multi-variable equation and isometric lines in the work of Du Carla, Halley, and Humboldt, as is obvious from the title of his report.[77]

Lalanne was not yet interested in exploring the cartographic implications of his work. Rather, he wanted to reduce the mathematical work of civil engineers. To simplify the construction of a graph, Lalanne proposed replacing one or more of the variables by some function of the variable and replacing arithmetic scales with logarithmic ones. He called this innovation anamorphic geometry and derived from it a device called the *abac*. Using a ruler (a special one was needed if more than three variables were involved), engineers could make calculations by determining the intersection of the graduated or calibrated lines representing the variables of the equation on a graph. Engineers of the Ponts et Chaussées used Lalanne's *abac* for decades, especially in determining the volumes of earth to be moved in constructing highways, canals, tunnels, and railroads.[78]

Lalanne pursued the resemblance between his mathematical work and mapping images in 1845 in a reaction to a report of one Morlet to the Academy of Science. In his report, Morlet related the problem of determining the center of a geometric shape to concepts of political economy.[79] Common references to the center of France or of Paris, Morlet believed, encouraged people to consider how administrative costs could be reduced if a unit were governed from its geographic center. To Morlet this approach was simplistic. He intended to determine the center of a figure in which several elements interact, such that the minimum spatial distance from any point to the center was relative to other values. The method he wished to apply involved making concentric lines of force, the distance between them representing a constant value for each variable; the point at which the different variables balanced each other would appear clearly. These circular lines he called *courbes d'égale excentricité*, because the sum of their distances to all the elements of a population was constant.

Lalanne's comments on Morlet's work were in turn published by the Academy. Lalanne generalized upon what Morlet had written and set it in the context of applied mathematics since the late eighteenth century. One long passage deserves to be quoted because it sets forth what an isopleth is (without giving it that name):

> Suppose, in effect, that the area of a country is divided into a very large number of units as small as France's communes; that at the center of each of these units a vertical line is raised that is in proportion to that unit's population, or in other words to the number of inhabitants per square kilometer ... ; that the ends of these vertical lines are connected by a curved surface, and that finally a map is made, at a convenient scale, and that numerical equivalents to these contour lines drawn on its surface are inscribed that correspond to equidistant integral elevations: one will thus have lines of equal specific population, from which the series of points where the population is thirty, forty, fifty, ... one hundred inhabitants per square kilometer will be noticeable. A map of this kind would offer as exact and vivid a representation of the distribution of population as possible. Similar to a topographical map drawn after the principles of Du Carla, it would show undulations, steep peaks, craters, passes, and valleys. One can foresee that its visible irregularity would be the reverse of that of the ground surface; thus our populous valleys would appear as mountain chains, while the barren heights of our mountains would appear as deep funnels. Such a map, if it existed, would render the greatest services in the study of many economic questions. It would supply ... the exact figure of persons served by a new transport line, and thus alleviate the contradictory results in matters of population distribution related to the adoption or rejection of proposed railway lines. Unfortunately, official government publications provide only area by arrondissement and population by canton, whereas to obtain truly satisfactory results it would be necessary to have these two figures by commune.[80]

Morlet's report, itself of interest for its cartographic implications, appears to have provoked Lalanne to articulate the concept of the isopleth as an abstract graphic representation of variables that do not actually exist in space at particular points.

Surprisingly, French engineers were slow to make isopleth maps.[81] The first isopleth map now known was the work of Nils Ravn, a Dane; it appeared in 1857 and depicted the population of Denmark. Probably the first French isopleth map was by Louis-Léger Vauthier, of the population of Paris, appearing in 1874.[82]

Common to Morlet, Lalanne, and Minard was a concern for the administrative implications of statistical analysis insofar as population, production, consumption, and resources are unevenly distributed in space. They enhanced the analytic potential of thematic mapping by generalizing its essential graphic expressions in logically consistent, quantitatively precise terms. That concern about political economy informed their work is no accident; engineers were expected to consider that topic. In 1818, the Ecole Polytechnique added a course in "social arithmetic" to prepare engineers to regulate and direct the execution of public works projects related to the expansion of cities and of industry. Engineers needed to evaluate the relative benefits and costs of competing proposals. Information on corporate finance, insurance, banking and related legal matters, general principles of probability, and demographic statistics (including the age and sex structure of the population, mortality rates, and life expectancy) were among the topics included in this course.[83] Charles-Joseph Minard (1781–1870) thought of adding a comparable but more advanced course at the Ecole des Ponts et Chaussées in 1831, but its introduction was delayed until 1847. By then, thematic mapping had emerged as a potent instrument of analysis. Minard advanced the state of the art by standardizing the design of graphic devices such as flow lines, pie charts, and proportional circles according to mathematically determined criteria.

Minard entered the Ecole Polytechnique at the age of sixteen and later graduated from the Ecole des Ponts et Chaussées.[84] His early work with the civil engineers involved canal and port development in Belgium and western France. From 1815 to 1822 he was assigned to Paris, and in the course of his duties he proposed a plan to supply paving stones to the city by canal and railroad.[85] Minard was one of the first Frenchmen to visit England to acquire firsthand knowledge of railroads. From 1822 to 1830 Minard supervised canal projects. From 1830 to 1836 he served as director of the Ecole des Ponts et Chaussées; he then stepped down for reasons of health but continued to teach at the school until 1842. In 1839 he was made a divisional inspector, and in 1846, in recognition of his work with thematic maps and of his understanding of transportation systems, he was appointed an inspector general and a member of the council of the Ponts et Chaussées. He retired from the council in 1851 when he reached the legally mandated retirement age, but he continued to serve on the editorial board of the *Annales des Ponts et Chaussées*, as he had since its creation in 1831; the *Annales* is the oldest modern professional civil engineering journal. Minard had a remarkable career. In his retirement, he was free to pursue interests he had lacked the time or encouragement for earlier. Although he made several important thematic maps before his retirement, a large proportion of his lifetime production came after 1851.

Minard's efforts to represent statistical data in map form began in the early 1840s. In two extraordinary reports published in 1842 and 1843, he demonstrated that the fares paid by long-distance travelers on a railroad do not adequately cover the costs of railroad operations. Passengers who travel between intermediate stations along the line, Minard believed, not only were more numerous, they were also more profitable for the railroad to carry. To make his point more convincing, he made a graph in 1844 showing the number of travelers carried annually between the various stations along the railroad lines. Robinson writes: "In his 'graphic table' he made the abscissa represent, according to scale, the length of the railroad line ... , while the ordinate was scaled according to the number of passengers traveling between the stations. By so doing he made the areas of the rectangles strictly proportional to the 'passenger kilometers' carried by the railroad. He shaded the portion which represented the passengers who travelled only [part of the way]."[86] He made one hundred copies of this chart.

Another person might have been content with charts, but Minard converted his chart into a thematic map, published by lithography in two hundred copies in 1845, for the council of the Ponts et Chaussées to use in evaluating proposed railway lines between Dijon and Mulhouse (fig. 28). On this map Minard showed the number of travelers using public stages on the various roads of the Franche-Comté by drawing flow lines whose widths were proportional to the traffic they represented. This map

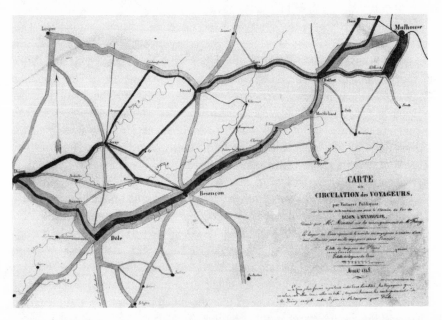

Figure 28 Traffic volume between Dijon and Mulhouse. This is one of the first maps by Charles Joseph Minard and one of the first made by anyone using flow lines. The lighter shades show the volume of through passengers using coaches on the highways; the dark shades show the volume of local traffic. The map was made to plan railroad routes. "Cartes de la circulation des voyageurs par voitures publiques sur les routes de la contrée où sera placé le chemin de fer de Dijon à Mulhouse." Printed by lithography (1845). 70 cm × 65 cm. Paris, Centre Pédagogique de Documentation et de Communication, Ecole Nationale des Ponts et Chaussées.

indicated that the railroad should be built in the valley of the Doubs River, and so it was. Minard made a map with flow lines showing road traffic between Lyon, Grenoble, and Valence in 1846. In 1845–46 other civil engineers made similar maps of traffic between Poitiers and La Rochelle and between Bordeaux and Bayonne. In 1854 the administration of the Ponts et Chaussées recommended the general use of Minard's method.

At the time he made his first thematic map, Minard's ideas about the genre were already fully formed. He perceived the potential uses of thematic maps long before he had an occasion to make and apply them. De Dainville and Robinson, who have resurrected Minard from the obscurity that has unjustly covered his work in this century, did not consider his maps in relation to his ideas about political economy. These ideas are worth examination. In 1850 Minard published a text on political economy.[87] He began by recounting that as early as 1831 he had proposed that a course in political economy be offered at the Ecole des Ponts et Chaussées. In that year he wrote a text showing how the concepts of political economy could be applied to the study of public works, but Minard delayed publication, for reasons he did not mention. Minard explained that his decision to publish was provoked by the opposition of several deputies in the Chamber in 1846 to the introduction of a course in political economy at the school the next year. Minard's text of 1850 was an enlarged version of his text of 1831.

Beginning with elementary notions of use value, capital, interest, and the like, Minard approached an understanding of the utility of public works. His view was that public works diminished the cost of transport, thereby simultaneously stimulating per capita consumption and enlarging the number of consumers. Utility should be measured in economic terms of cost and prices, so that the contribution of dissimilar structures such as railroads and canals could be compared. Minard attributed to canals a competitive advantage in long-distance trade, observing that very little freight is carried between two intermediate points on a canal. The advantages of the railroad were in time saved and in service to intermediate points. The benefits of competition between canal and railroad, Minard believed, would outweigh the costs. The relation between Minard's ideas of the 1830s and his use of charts and maps in the 1840s are obvious.

The most important issue Minard addressed in his text was that of choosing between different projects for public works. He began with two observations: first, that forecasts of future traffic are subject to error, and second, that public works are inherently temporary despite their solid and permanent appearance. Elaborating upon his second point, Minard argued that too much time, money, and capital are invested in public works when they are designed to last a long time. Much better, Minard believed, to invest a lot in a few projects that advance rapidly than to spend modest sums on many projects that take longer to complete. Given migration of people, the relocation of industry, and the rate of economic change, it is easier, cheaper, and more prudent to rebuild the infrastructure every generation than to build it to last for several generations. He concluded that there is no perpetual utility.

From these observations, certain consequences for the selection and funding of public works followed. Given that the state cannot support all worthwhile public works projects, how can it determine which are of national merit and which only of local interest? How can the risk involved in a given project be assessed? How should the routes for canals and railroads be selected? It was by no means clear to Minard that a centralized state bureaucracy was better equipped to make these decisions than syndicates of investors operating in the marketplace.

Many of Minard's thematic maps take on new meaning in this context, from his early map of the Dijon-Mulhouse route to his maps of English coal exports and of European cotton imports in 1866. His maps were not merely sophisticated graphic representations of statistics extracted from dozens of tables and scores of monographs. Minard argued that his maps did not speak to the eye, but counted.[88] His flow lines and pie charts were designed so that size was strictly proportional to the statistical values represented. Minard intended his work to place the decisions of the council of the Ponts et Chaussées on a more rational, systematic, cost-effective basis, so that France's transportation infrastructure would be adequate to meet the needs of its cities and industries. At a time of rapid and unpredictable economic change, Minard's maps brought basic patterns to light and described France's place in the evolving world economy. From 1851 on, he made nearly fifty maps for the government; Minard estimated that the total number of copies approached ten thousand.

Minard perfected and popularized a variety of techniques to represent data clearly and accurately. In the process, he subordinated geographic accuracy to statistical accuracy. That Minard distorted the shape of continents or the proportions of the seas matters less than the accuracy with which he drew the width of flow lines or the size and division of pie charts.[89] His subjects included most commodities, but on two occasions he also took on historical subjects; in 1869 Minard used flow lines to show reductions in the sizes of the armies of Hannibal in his campaign into Italy and of Napoleon in retreat from Russia. Most of the maps were economic: livestock carried to Paris by railway in 1862, with four colors to distinguish sheep, calves, pigs, and cattle and flow lines calculated to reflect the different weight per animal of these four kinds; passenger traffic on main European railroads; canal freight tonnage for various years; freight traffic in French ports and in the major ports of the world; and the import and export of various commodities by different means of transport, such as coal, wine, and cotton. A map of coal freight shipments in France in 1845, made in 1851, used colors to distinguish coal mined in France in three locations as well as German, Belgian, and British imports, and used circles to show the percentage of French-mined coal consumed in the department of production (plate 8). A map of 1858 showed global population migration patterns. Many of the maps treated the same subject annually so that two or three maps, placed side by side, made changes over time apparent. One very unusual map was made in 1865 to determine where the new central post office should be built in Paris. Minard thought the post office should be centrally located, a little to the west of the geographic center of the population to be closer to government offices that generated so much mail and to the railroad stations. He calculated that the demographic center of Paris was the southwest corner of Les Halles; it had been at Saint Eustache nearby in 1831 and had moved 150 meters in thirty years. The point in Paris equidistant from the railroad stations as a function of postal freight by weight Minard calculated to be at the intersection of the rue Beaubourg and the rue du Maure. Minard made the map with symbols to represent the ministries, black squares to show the population size of Paris's districts, and shaded squares to show the weight of mail generated per district.

That Minard's methods were followed by his successors is perhaps the best measure of their effectiveness.[90] From 1872, the Direction des Chemins de Fer produced flow maps of the traffic revenue of the railroads. The Ministry of Public Works established a map division in 1877 under the direction of J. J. Emile Cheysson (1836–1910) to prepare thematic maps and charts illustrating statistical information graphically. Cheysson was concerned with two interrelated problems in thematic cartography: comparisons between dissimilar topics, such as medical and agricultural data, and the elimination of irrelevant and excessively heterogeneous information. Minard would have been satisfied to know that seventeen albums of thematic maps were prepared for the Chamber of Deputies between 1879 and 1899. And no doubt he would have had something to say when budgetary considerations brought the series to an end in 1900. But the work of Cheysson belongs to another period, one which involves many topics that have yet to be studied as intensively as cartography in more remote times.

This book began at a decisive point in the development of cartography, when the leadership in cartography passed from the Netherlands to France and when scientific principles supplanted a pictorial tradition. It ends less abruptly. The accomplishments of the French in geodetic, topographical, and thematic mapping enlarged the scope of cartography and the influence of the state in ways that other nations emulated. In France more than elsewhere between 1660 and 1848, conditions were favorable to the development of cartography. Cartography became an autonomous enterprise; although its development was profoundly conditioned by the ideas and careers of certain individuals and by the evolution of certain institutions, so that the fate of the national map survey or the contour line might be in question for years at a time, ultimately enough people and institutions became involved in cartography to sustain its development.

From this perspective, it seems that the contribution of the French state to cartography was greater than the contribution of cartography to the state. Canals and roads could have been constructed, and the frontiers defended, had the French possessed only inferior maps. But state support for cartography was not evaluated by contemporaries according to a narrow calculation of costs and benefits. Although the state supported cartography in the expectation that better maps and map collections would promote internal improvements and commerce and strengthen foreign policy and defense, the impulse within government to support cartography was not limited to short-term objectives or specific projects. What characterized the relation between the state and cartography was the willingness of government to make extensive mapping surveys and large map collections a routine part of government, so that maps would be available in advance of need. On this scale, the mapping of a nation corresponded to the same rational purpose as the mapping of an estate: geodetic measurement, the cartographic representation of landforms, and the synthesis of statistics in thematic maps were a way of claiming possession, a means of asserting knowledge as an instrument of control.

As cartography evolved in France, the technical aspects of making maps became more complex and sophisticated, trends that encouraged the establishment of bureaucratic units for cartography. The government did not monopolize cartography; on the contrary, it conceived for itself the role of diffusing improved maps and mapping standards for the benefit of the public. A gain in rigor and precision might have discouraged the interest of people whose understanding of cartographic methods was elementary and superficial, but the humanistic tradition of the Renaissance survived the modernization of cartography in the eighteenth century. As maps became more sophisticated in France, a generalization of map usage occurred. This established a paradigm for the role of maps in modern culture that the history of cartography can illuminate. In the twentieth century, when the camera, airplane, and computer have simultaneously increased the complexity of cartography and broadened its influence, humanists are again challenged to imagine new uses for maps, and politicians are again confronted with the task of deciding how to support cartography. Maps and mapping remain metaphors for knowledge and its acquisition. From the perspective of the present, the role of maps in modern culture appears to rest solidly on the achievements of French scientists, engineers, and public administrators active from the middle of the seventeenth century to the middle of the nineteenth.

Notes

Abbreviations Used in Notes

AAE	Archives du Ministère des Affaires Etrangères, Paris
AN	Archives Nationales, Paris
BN	Bibliothèque Nationale, Paris
IGN	Cartothèque, Institut Géographique National, Saint-Mandé
MARS	*Mémoires de l'Académie Royale des Sciences*
PC	Ecole Nationale des Ponts et Chaussées, Paris
SHAT	Service Historique de l'Armée de Terre, Vincennes

Preface

1. C. Sandler, *Die Reformation der Kartographie um 1700* (Munich: Oldenburg, 1905).

Introduction

1. Rhoda Rapoport, "Government Patronage in Science in Eighteenth-Century France," *History of Science* 8(1961):122. Rapoport provided a list of questions for further study: "To what extent and for what purposes did government officials recruit scientific talent? . . . How did officials distinguish between men of ability and the less talented, between the practicable scheme and the technically less feasible one? . . . Can one discern particular programs and policies, or did officials generally respond to the needs, projects and suggestions of the moment?" (124).

2. Charles C. Gillispie, *Science and Polity in France at the End of the Old Regime* (Princeton: Princeton University Press, 1980).

3. The best introductions to the history of cartography as a field are M. J. Blakemore and J. B. Harley, "Concepts in the History of Cartography: A Review and Perspective," Monograph 26, *Cartographica* 17(1980):1–120; R. A. Skelton, *Maps: A Historical Survey of Their Study and Collecting* (Chicago: University of Chicago Press, 1972); Arthur H. Robinson and Barbara Bartz Petchenik, *The Nature of Maps: Essays toward Understanding Maps and Mapping* (Chicago: University of Chicago Press, 1976); and David Woodward, "The Study of the History of Cartography: A Suggested Framework," *American Cartographer* 1(1974):101–15.

4. Mary Sponberg Pedley, "The Map Trade in Paris, 1650–1825," *Imago Mundi* 33(1981):33–45; idem, "The Subscription List of the 1757 'Atlas Universel': A Study in Cartographic Dissemination," *Imago Mundi* 31(1979):66–77; François de Dainville, *Le langage des géographes* (Paris: Picard, 1964); Mireille Pastoureau, "Les atlas imprimés en France avant 1700," *Imago Mundi* 32(1980):45–72; idem, *Les atlas français XVIᵉ–XVIIᵉ siècles: répertoire bibliographique et étude* (Paris: Bibliothèque Nationale, 1984).

1. Skelton, Maps, 76–78.
2. Robinson and Petchenik, Nature of Maps, 16–20.
3. De Dainville, Langage des géographes, 59.

Chapter One

1. David Buisseret, "Les ingénieurs du roi au temps de Henri IV," Ministère de l'Education Nationale, Comité des Travaux Historiques et Scientifiques, Bulletin de la Section de Géographie, 1964 (1965), 13–81. An interesting comparison can be drawn between England and France from the article by Victor Morgan, "The Cartographic Image of 'the Country' in Early Modern England," Transactions of the Royal Historical Society, ser. 5, 29(1979):129–54.

2. Jean-Baptiste Colbert, "Instruction pour les maîtres des requêtes, commissaires départies dans les provinces," September 1663. Reprinted in Pierre Clément, ed., Lettres, instructions et mémoires de Colbert (Paris: Imprimerie Nationale, 1861–73), 4:27; see also 5:442.

3. C. A. Crommelin, Physics and the Art of Instrument Making at Leyden in the Seventeenth and Eighteenth Centuries (Leiden, 1926); J. Keuning, "Sixteenth Century Cartography in the Netherlands," Imago Mundi 9(1952):35–65; E. G. R. Taylor, "The Measure of the Degree: 300 B.C.–A.D. 1700," Geography 34(1949):121–31; idem, "The Earliest Account of Triangulation," Scottish Geographical Magazine 43(1942):341–45; A. Pogo, "Gemma Frisius: His Method of Determining Difference of Longitudes by Transporting Timepieces (1530) and His Treatise on Triangulation (1533)," Isis 22(1935):469–85; N. D. Haasbroek, Gemma Frisius, Tycho Brahe and Snellius and Their Triangulations (Delft: Netherlands Geodetic Commission, 1968). According to Haasbroek, p. 10, thirty editions of Frisius's Cosmographicis liber Petri Apiani were published between 1524 and 1609, including sixteen in Latin, eight in Dutch, and five in French; twenty-nine of these carried an appendix describing triangulation.

4. D. W. Waters, The Art of Navigation in England in Elizabethan and Early Stuart Times (New Haven: Yale University Press, 1958); A. W. Richeson, English Land Measuring to 1800: Instruments and Practices (Cambridge: MIT Press, 1966), 100–03; E. G. R. Taylor, The Haven-Finding Art: A History of Navigation from Odysseus to Captain Cook (London: Hollis and Carter, 1956); Norman J. W. Thrower, ed., The Compleat Plattmaker: Essays on Chart, Map and Globe Making in England in the Seventeenth and Eighteenth Centuries (Berkeley and Los Angeles: University of California Press, 1978).

5. Norman J. W. Thrower, "Edmond Halley and Thematic Geo-Cartography," in Compleat Plattmaker; Angus Armitage, Edmond Halley (London: Nelson, 1966); Colin A. Ronan, Edmond Halley: Genius in Eclipse (New York: Doubleday, 1969). See also the valuable background on the idea of maps of magnetic variation in the seventeenth century by Helen Wallis, "Maps as a Medium of Scientific Communication," in Studia z Dziejow Geografii i Karbografii: Etudes d'histoire de la géographie et de la cartographie, ed. Jozef Babicz, (Warsaw: Polska Akademia Nauk Zakkad Historii Nauki i Techniki, 1973), 251–62.

6. Roger Hahn, The Anatomy of a Scientific Institution: The Paris Academy of Sciences, 1666–1803 (Berkeley and Los Angeles: University of California Press, 1971), 2.

7. Derek Howse, Greenwich Mean Time and the Discovery of Longitude (Oxford: Oxford University Press, 1980).

8. René Taton, "Gian Domenico (Jean-Dominique) Cassini," Dictionary of Scientific Biography, 3(1971): 100–104.

9. Léon Gallois, "L'Académie des Sciences et les origines de la carte de Cassini," Annales de Géographie 18(1909):194–204, 289–310.

10. Juliette Taton and René Taton, "Jean Picard," Dictionary of Scientific Biography, 10(1974):595–97.

11. Jean Picard, La mesure de la terre (Paris: Imprimerie Royale, 1671).

12. J. W. Olmsted, "The Scientific Expedition of Jean Richer to Cayenne (1672–1673)," Isis 34(1942):117–28. Simultaneous observations by Richer in America and Cassini in Paris, and by Cassini in Paris and Picard in Denmark, enabled astronomers to determine the parallax of the sun and Mars and to estimate more accurately the mean distance between the earth and the sun and the dimensions of planetary orbits.

13. Ibid., 128.

14. Gallois, "Académie des Sciences," 289–92.

15. "Route de Paris à Nantes" (1679), Manuscript map probably by La Hire, Service Historique de la Marine, receuil SH 5:26.

16. "Carte de France corrigée par ordre du Roi sur les observations de Mrs de l'Académie des Sciences." Paris, undated, BN, Cartes et Plans, Ge. DD. 2987–777. Projects by the hydrographer Chazelles from 1687 for revising Mediterranean charts are in AN, Marine 3 JJ 185.

17. An unpublished diary by Cassini I from the last two years of his life, dictated when he became blind, reveals that at that time he was far more concerned with his family and with religious studies and ideas than with scientific problems. "Journal de la vie privée de Jean-Dominique Cassini dans les deux dernières années de sa vie depuis le 1er juin 1710 jusqu'au 11 septembre 1712 dicté par lui-même jusqu'au moment de sa mort." BN, Cartes et Plans, Ge. DD. 2066.

18. Jacques Cassini, *De la grandeur et la figure de la terre (Suite de Mémoires de l'Académie royale des sciences, 1718)* (Paris, 1720). The maps made of the triangulation include "Carte fort curieuse de la Meridienne d l'Observatoire de Paris" (1718), 504 cm × 46 cm, and "Carte des provinces de France traversées par la Meridienne de Paris contentant les . . . triangles qui ont servi à en déterminer la partie septentrionale astronomiquement et géometriquement," seven sheets, each 47 cm × 61 cm. BN, Cartes et Plans, Ge. DD. 5509 and 5510, respectively. On these maps the triangles of the primary network were in red and those used in verifying the first network were in dotted lines; the names of places accurately determined by astronomic and geodetic observations were underlined, those of places determined by only one method were underlined with dots, and those of unverified places were not underlined at all.

19. The only source of this information is Cassini I's journal cited in note 17. Cassini recorded his daughter-in-law's progress in pregnancy but failed to make an entry during a ten-day period when the child was born.

20. Eugène-Jean-Marie Vignon, *Etudes historiques sur l'administration des voies publiques en France aux XVIIe et XVIIIe siècles*, 4 vols. (Paris: Dunod, 1862); Henri Marie Auguste Berthaut, *Les ingénieurs géographes militaires, 1624–1831: Etude historique*, 2 vols. (Paris: Service Géographique de l'Armée, 1902); Jean Petot, *Histoire de l'administration des Ponts et Chaussées (1599–1815)* (Paris: Marcel Rivière, 1958); Anne Blanchard, *Les ingénieurs du "roy" de Louis XIV à Louis XVI: Etude du corps des fortifications*, Collection du Centre d'Histoire Militaire et d'Etudes de Défense Nationale de Montpellier, 1979). The orders establishing the navy's hydrographic office are in AN, Marine 1 JJ 1.

21. Harcourt Brown, "From London to Lapland and Berlin," in *Science and the Human Comedy: Natural Philosophy in French Literature from Rabelais to Maupertuis*, 167–206 (Toronto and Buffalo: University of Toronto Press, 1976), 169.

22. Letters by Cassini II to Orry, 24 October 1733, 30 May 1735, 20 July 1735, and 20 August 1735, BN, Cartes et Plans, Rés. Ge. DD. 3206; one of the accompanying maps is Rés. Ge. C. 9987.

23. Brown, "From London to Lapland and Berlin"; Florence Trystram, *Le procès des étoiles* (Paris: Seghers, 1979).

24. SHAT A^{1}3011, fols. 133, 147, 216, 221, 304.

25. Buache memorandum, BN, Cartes et Plans, Ge. FF. 13732–II; model album of 116 sheets, BN, Département des Estampes, petit in-fol. Ve 6.

26. Guy Arbellot, "La grande mutation des routes de France au milieu du XVIIIe siècle," *Annales Economies, Sociétés, Civilisations* 28(1973):765–90.

27. Gallois, "Origines de la carte de Cassini," 305–7.

28. The full title of Cassini III's unpublished manuscript is "Le Parfait Ingénieur: Ouvrage dans lequel on expose avec le plus grand détail la théorie et la pratique de l'art de lever les plans en employant tout ce que l'Astronomie et la Géométrie offrent de secours pour la perfection de la géographie." Archives of the Observatory of Paris, no. D2–44. Cassini's speech to the Academy of 13 November 1745 is quoted by Gallois, "Origines de la carte de Cassini," 304. Josef W. Konvitz, "Redating and Rethinking the Cassini Geodetic Surveys of France, 1730–1750," *Cartographica* 18(1982):1–15; idem, "The National Map Survey in Eighteenth-Century France," *Government Publications Review* 10(1983):395–403.

29. François Chevalier, "Sur une manière de lever la carte d'un pays," *Histoire de l'Académie Royale des Sciences*, 1707(1708):113–18. Jean Baptiste d'Anville, "Mémoire Instructif pour que dans toutes les Paroisses d'un dioèse, il soit dressé en même-tems & uniformément, par une méthode

aisée à pratiquer, des Cartes et des Mémoires particuliers, qui puissent fournir un détail suffisant pour la carte générale de ce diocèse ou d'une province" (1732); idem, "Mémoire instructif pour dresser sur les lieux des cartes particulières et Topographiques d'un Canton de Pays, renfermant dix ou douze Paroisses" (1743). The former can be found in the collection of printed material ("imprimés") of the Bibliothèque Nationale under the number Vp 527, the latter under the number Vp 525. Another pamphlet by d'Anville, number Vp 521, was published for the intendant of the généralité of Soissons in 1745. There is a blank base map in BN, Cartes et Plans, Ge. D. 10820.

30. C. Lemoine-Isabeau, "Lettres d'ingénieurs géographes français en Flandres, en 1746," *Revue Belge d'Histoire Militaire* 13(1980):413–26.

31. Henri Marie Auguste Berthaut, *La carte de France, 1750–1898: Etude historique*, 2 vols. (Paris: Service Géographique de l'Armée, 1898), remains an authoritative treatment.

32. "Mémoires de Jean-Dominique Cassini IV," BN, Cartes et Plans, Ge. DD. 2066; Cassini IV's published autobiography is *Mémoires pour servir à l'histoire des sciences et à celle de l'Observatoire Royal de Paris* (Paris: Bleuet, 1810). The principal secondary source about Cassini IV is Jean-François Schlisteur Devic, *Histoire de la vie et des travaux scientifiques et littéraires de J. D. Cassini IV* (Clermont, 1851).

The Paris Observatory has a series of reports and letters relating to the Greenwich-Paris operation of 1787 in its archive, D. 57. See also Cassini IV, Méchain, and Le Gendre, *Exposé des opérations faites en France en 1787 pour la jonction des observatoires de Paris et de Greenwich* (Paris, 1790). William Roy published "An Account of the Mode proposed to be followed in determining the relative Situation of the Royal Observatories of Greenwich and Paris," Royal Society of London, *Philosophical Transactions* 77(1787):133–226, and "An Account of the Trigonometrical Operation, whereby the distance between the meridians of the Royal Observatories of Greenwich and Paris has been determined," Royal Society of London, *Philosophical Transactions* 80(1790):111–270. See also Sir Charles Close, *The Early Years of the Ordnance Survey* (1926; reprinted, ed. John Brian Harley, New York: Augustus M. Kelley, 1969), 12–14, and R. A. Skelton, "The Origins of the Ordnance Survey of Great Britain," *Geographical Journal* 128(1962):415–30.

33. Gillispie, *Science and Polity*, 126. Gillispie provides a clear description of the design and use of Borda's repeating circle, 127–30. Gillispie also gives the full references to Legendre's publications.

34. Richeson, *English Land Measuring to 1800*, 169–70.

35. Maurice Daumas, *Les instruments scientifiques aux XVIIe et XVIIIe siècles* (Paris: Presses Universitaires de France, 1953), 135–43. See also Gillispie, *Science and Polity*, 118–24, for a review of French attempts at reforming instrument making at the end of the Old Regime; the quotation is from p. 122.

36. François de Dainville, *La carte de la Guyenne par Belleyme, 1761–1840* (Bordeaux: Delmas, 1957).

Chapter Two

1. Roger Dion, *Les frontières de la France* (Paris: Hachette, 1947); Nelly Girard d'Albissin, *Genèse de la frontière franco-belge: Les variations des limites septentrionales de la France de 1659 à 1789*, Bibliothèque de la Société d'Histoire du Droit du Pays Flamands, Picards, et Wallons, no. 26 (Paris: Picard, 1970); Norman J. G. Pounds, "The Origin of the Idea of Natural Frontiers in France," *Annals of the Association of American Geographers* 41(1951):146–57; John R. Stilgoe, "Jack o' Lanterns to Surveyors: The Secularization of Landscape Boundaries," *Environmental Review* 1(1976):14–31; David Buisseret, "Cartography and Power in the Seventeenth Century," *Proceedings of the Tenth Annual Meeting of the Western Society for French History* 10(1984):103–5; P. D. A. Harvey, *History of Topographical Maps: Symbols, Pictures and Surveys* (London: Thames and Hudson, 1980), 88–103, 158; Monique Pelletier, "La Martinique et la Guadeloupe au lendemain du Traité de Paris (10 fevrier 1763): L'oeuvre des ingénieurs géographes," *Chronique d'Histoire Maritime*, no. 9(1984):22–30.

2. G. N. Clark, *The Seventeenth Century* (Oxford: Clarendon Press, 1929; reprinted Oxford University Press, 1960), 144. A representative map is "Carte des limites de France et des terres cédées depuis la mer jusqu'à la Lis avec les enclavements desdittes terres cédées dans celles de France et le chemin projetté depuis Armentières jusques Bergues par Bailleul et Steinwurde," 1714. 1:57,600; 93 cm × 52 cm. Archives du Génie (Vincennes), art. 4, sect. 3, para. 1, carton 1, no. 3.

3. Chevalier de Bonneval, "Mémoire pour l'establissement général des limites du royaume," 29 December 1747. Archives du Génie (Vincennes), art. 4, sect. 3, para. 1, carton 1, no. 6 bis.

4. "Etat des cartes géographiques du Roy," 1776. BN, Cartes et Plans, Ge. FF. 13427.

5. Paul Poindron, "Les cartes géographiques du Ministère des Affaires Etrangères (1780–1789): Jean-Denis Barbié du Bocage et la collection d'Anville," *Sources, Etudes, Recherches, Informations des Bibliothèques Nationales de France*, ser. 1, no. 1(1943):46–72.

6. Letters by Barbié du Bocage, BN, Cartes et Plans, Ge. FF. 15631.

7. The Bibliothèque Nationale has other items from the original d'Anville collection that the foreign office did not acquire. D'Anville's manuscript notes on units of measurement, on geography, and on the history of geography went directly into the Bibliothèque Nationale. D'Anville's books and about one hundred manuscript maps entered the possession of Louis-Charles-Joseph de Manne, who began his career at the library in 1791 and became curator of books there in 1820. His widow eventually admitted she had these items but refused to surrender them. They were dispersed at a sale in 1863. Edmond de Manne, son of Louis and also a librarian, sold about 1,000 maps, including some by d'Anville, to the library. Barbié du Bocage, for his part, owned 2,178 maps, including about 80 by d'Anville, that went on sale in 1844. Most of these were acquired for the library by Edme Jomard.

8. Jean-Pierre Samoyault, *Les Bureaux du Secrétariat d'Etat des Affaires Etrangères sous Louis XIV*, Bibliothèque de la Revue d'Histoire Diplomatique, no. 3 (Paris: A. Pedone, 1971), 152–54; AAE, Personnel, ser. 1, vol. 63, entry for Gaillard de Saudray. See also "Projet d'un bureau et dépôt géographique au Département des Affaires Etrangères," an undated document in which the point was made that maps are useful in making pacts, formulating strategies, exchanging land, and examining proposals by and pretensions of other powers. BN, Cartes et Plans, Ge. FF. 13291.

9. AAE, Personnel, ser. 1, vol. 69, entry for Rizzi-Zannoni; Ludovic Drapeyron, "J. A. Rizzi-Zannoni: Son séjour en France," *Revue de Géographie* 39(1897):401–13; Berthaut, *Ingénieurs géographes*, 1:218–19.

10. AAE, Rizzi-Zannoni, fol. 259.

11. Ibid., fol. 280.

12. AAE, Personnel, ser. 1, vol. 36, entry for Grandjean, fol. 59, 60.

13. AAE, Grandjean, fol. 66, 69, 71.

14. Ibid., fol. 87, "Observations sur la levée d'une carte topographique," 28 April 1789, and fol. 89, "Projet d'instructions pour la levée de la carte topographique de la frontière des Pyrénées . . . en 1786 . . . ," undated.

15. "Nouvelles dispositions pour le travail de la levée de la carte de la frontière des Pyrénées relativement à l'etablissement de la limite entre la France et l'Espagne," 17 February 1789, SHAT, Correspondence topographique, A. 2; Berthaut, *Ingénieurs géographes*, 2:103–5.

16. AAE, Grandjean, fol. 98, 103, 116.

17. Berthaut, *Ingénieurs géographes*, 2:401–4, 438–41. At the IGN there are two manuscript maps of the boundaries of France made as part of the treaty-making process at the end of the Napoleonic Wars: "Carte de la frontière d'entre Rhin et Moselle indiquant les limites de cette frontière en 1790 et d'après les traités du 1814 et 1815, assemblage de la Carte de Cassini au 1:86,400, de Thionville à Lauterbourg (1815)," no. 160; and "Carte de la limite des frontières du Nord de la France d'après le traité de 1815 . . ." (1819 and after), no. 162.

18. Léon Desbuissons, "Exposé historique" (on the Bureau des Limites), undated, ca. 1906–7, consulted in the reading room of the library, Ministère des Affaires Étrangères.

19. In 1871 cadastral maps were used in determining the Franco-German boundary. Aimé Laussedat, *La délimitation de la frontière franco-allemande* (Paris: Ch. Delagrave, 1901), includes a full discussion of how maps were used and several maps showing proposed boundary lines.

20. "Projet de travail à faire faire pendant la paix par les ingénieurs géographes . . . proposé par le Sr. Berthier, à M. le Duc de Choiseul en 1762 . . . ," BN, Cartes et Plans, Ge. FF. 13292. Berthier recommended making the "Carte des chasses," the famous, beautifully engraved map of the royal hunting grounds near Paris, as a training program for army engineers. To keep engineers employed during peace, he also recommended making maps of the coasts of France and of French colonies, of France's inland frontiers, and of the itineraries and battle formations of armies in the past.

21. D'Arçon, "Disposition générale du travail . . . de la carte du Dauphiné et de la Provence . . . ," Archives du Génie (Vincennes), art. 4, sect. 1, para. 5[10], carton 2, no. 1[6].

22. D'Arçon, "Instruction sur l'objet des opérations topographiques qui ont été ordonnées en haute Alsace . . . ," Archives du Génie (Vincennes), art. 4, sect. 1, para. 3, carton 3, no. 16.

23. D'Arçon, "Réflexions sur l'exécution de la topographie des frontières . . . ," Bibliothéque du Génie (Paris), in-fol. MS. 210.

24. On de Bourcet, see Basil Liddell-Hart, The Ghost of Napoleon (New Haven: Yale University Press, 1935). De Bourcet has not yet been the subject of a monograph, yet the length and importance of his career and the volume of his written and cartographic materials could easily support one.

25. Gillispie, Science and Polity, 198.

26. Gillispie, Science and Polity, provides the most recent study of science and reform before 1789. Historians of science have tended to emphasize the political role of scientists and political debates about scientific institutions and education rather than the actual scientific work conducted during the Revolution. An exception is Carl B. Boyer, "Mathematicians of the French Revolution," Scripta Mathematica 25(1960):11–31. See also Joseph Fayet, La révolution française et la science, 1789–1795 (Paris: Marcel Rivière, 1960); Hahn, Anatomy of a Scientific Institution; Henry Guerlac, "Some Aspects of Science during the French Revolution," Scientific Monthly 80(1955):93–101; Guerlac's observation, that the output of scientists was more distinguished and of greater significance than what artists or writers produced, does not appear to have been pursued further.

27. AN, K 879; M. Fougères, "Les plans cadastraux de l'Ancien Régime," Mélanges d'Histoire Sociale, Annales d'Histoire Sociale 3(1943):54–69.

28. Paul Guichonnet, "Le cadastre savoyard de 1738 et son utilisation pour les recherches d'histoire et de géographie sociales," Revue de Géographie Alpine 43(1955):255–98; Musée Savoisien, Le cadastre sarde de 1730 en Savoie (Chambéry, 1980).

29. Antoine Albitreccia, Le plan terrier de la Corse au 18ᵉ siècle: Etude d'un document géographique (Paris: Presses Universitaires de France, 1942); Marcel Huguenin, "La cartographie ancienne de la Corse," Bulletin d'Information de l'Association des Ingénieurs géographes, no. 23 (July 1962):85–98; no. 26 (July 1963):33–55; idem, "French Cartography of Corsica," Imago Mundi 24(1970):123–37. In 1791 the cadaster was nearly complete. The National Assembly authorized supplementary funds of 38,234 livres to finish the job, which had already cost 250,000 livres. In 1794 Corsica fell to the English, who nevertheless allowed the French to finish on condition that the English be given all the records. The French were allowed to make copies of the maps, which were placed in the Sorbonne. Prony later copied them for Napoleon. In the year VI, Lalande and Monge recommended that the map be engraved at a scale of 1:86,400 (the same as the second Cassini national map survey) and that a final notebook be published recording the past and present condition of the island, together with changes recommended for it. Disputes between the codirectors Testevuide and Bédigis and their heirs and the government over whether the state still owed payments for the cadaster lasted intermittently until 1810. The map was published in eight sheets in 1824.

30. M. D. T. D. V. [Dutillet de Villars], Précis d'un projet d'etablissement du cadastre dans le royaume (Paris: Imprimerie de Clousier, 1781), and "Prospectus pour l'établissement du cadastre dans tout le royaume," 27 July 1778, BN, Manuscrits, Fond Français, MS. 11217.

31. F. N. Babeuf and J. P. Audiffred, Cadastre perpétuel (Paris: Garnery et Volland, 1789), 52; Munier, Essai d'une méthode générale propre à étendre les connaissances, ou Recueil d'observations, 2 vols. (Paris: Moutard, 1779), 1:196–97.

32. The original maps and "cahiers" of the communities of France are in the map division of the Archives Nationales, NN *9–14. See also Pierre Delaunay, "Un projet de division géométrique du territoire français à la fin du 18ᵉ siècle," Bulletin de la Librairie Ancienne et Moderne, no. 121 (January 1970):2–6; René Faille, "La carte de France divisée en carrés par Robert de Hesseln," Bulletin de la Librairie Ancienne et Moderne, no. 127 (August–September 1970):122–27; Georges Mage, La division de la France en départements (Toulouse: Imprimerie Saint-Michel, 1924); and Numa Broc, La géographie des philosophes: Géographes et voyageurs au 18ᵉ siècle (Paris: Orphys, 1975), 460–65.

33. Georg Strasser, "The Toise, the Yard and the Metre—The Struggle for a Universal Unit of Length," *Surveying and Mapping* 35(1975):25–46.

34. Fayet, *Révolution et science*, 444–48; C. Stewart Gillmor, *Coulomb and the Evolution of Physics and Engineering in Eighteenth-Century France* (Princeton: Princeton University Press, 1971), 70–73.

35. Charles Coulston Gillispie, "Laplace," *Dictionary of Scientific Biography*, 15(1978):333–35.

36. Maurice P. Crosland, *Science in France in the Revolutionary Era* (Cambridge: MIT Press, 1969), 195; idem, " 'Nature' and Measurement in Eighteenth-Century France," *Studies on Voltaire and the Eighteenth Century* 87(1972):277–309.

37. Gillispie, "Laplace," 335.

38. See, for example, the comments of Delambre, in PC, MS. 726.

39. Robert McKeon, "Gaspard-François de Prony," *Dictionary of Scientific Biography*, 11(1975):163–66.

40. Prony, "Réflexions sur la carte et le cadastre de la France," 10 October 1791, PC, MS. 2147.

41. Bibliothèque de l'Institut, MS. 2317, no. 13.

42. Prony, "Exposé des travaux faits par le bureau du cadastre de la France, depuis son éstablissement au mois d'octobre 1791 jusqu'au 20 mai 1792," 31 May 1792, PC, MS. 2148; idem, "Situation de travail des Bureaux du cadastre et transports au 30 Frimaire An II," PC, MS. 2402; idem, "Rapport sur le travail de la 3ᵉ division de l'agence des cartes et plans," 26 Messidor An II, PC, MS. 2402.

43. Prony's tables computed "logarithms of sines and tangents to fourteen decimals in tens of centesimal 'seconds' and logarithms of numbers from 1 to 100,000 to nineteen decimals." I. Bernard Cohen, "Jean-Joseph Delambre," *Dictionary of Scientific Biography*, 4(1971):16. See also Lagrange, Laplace, and Delambre, *Notice sur les grandes tables logarithmiques et trigonometriques calculées au Bureau du Cadastre* (Paris: Baudouin, 1801).

44. "De la quantité et du format du papier qui doit être employé à la confection du cadastre du royaume," PC, MS. 2150.

45. "Aperçu des dépenses nécessaires à la continuation des travaux astronomiques et géographiques du Bureau du cadastre . . . ," PC, MS. 2148; "Mémoire sur la nécessité de former une Division du Bureau du Cadastre [qui doit s'occuper] de l'instruction des ingénieurs géographes," August 1793, PC, MS. 2402.

46. Owen Gingerich, "Pierre-François André Méchain," *Dictionary of Scientific Biography* 9(1974):250–52; Joseph Laissus, "Un astronome français en Espagne: Pierre-François-André Méchain," in 94th Congrès National des Sociétés Savantes, Paris 1969, *Comptes-Rendues, Sciences* (Paris: Bibliothèque Nationale, 1970), 1:37–59; Jean-Joseph Delambre, *Base du système métrique decimal*, 3 vols. (Paris, 1806, 1807, 1810).

47. Meanwhile Méchain yearned to return to Spain. In 1803 he tried to extend his observations to the Balearic Islands as a check on his earlier calculations, but when he reached Ibiza he realized he could not see the mainland. Exhausted and ill from yellow fever, he died in 1804 while working south of Barcelona. Jean-Baptiste Biot and François Arago completed his work in 1806–8.

Delambre's measurements were accurate to one part in 36,000, or a difference of less than three-tenths of a meter along the meridian from Perpignan to Dunkerque. When Delambre examined Méchain's notes after the latter's death as part of his final report, he discovered that an error of three seconds had been the occasion of Méchain's obsession. In 1810 the Institut awarded Delambre a prize for his share in the measurement of the meridian.

48. "Résultats du travail fait au bureau du cadastre pour connoitre la superficie et la population du territoire français" (1795?), PC, MS. 2148. After subtracting uncultivable land from all land using a formula of Lavoisier's (which expressed this difference as the ratio of 648/1,050), Prony estimated the density of France, taking only cultivable land into account, at one person per 1.328 hectares. "Tableau figuré de la France contenant sa division en départements, sa population et sa superficie, d'après les grandes tables de population et toisé général du territoire français fait au bureau du cadastre" (1796) made this information available to the public.

49. See Crosland, *Science in France*, 17, 50–51, for the observations of the Danish astronomer Thomas Bugge about the school in 1798.

50. Miscellaneous reports and letters between Prony and the administrators of the Ponts et Chaussées, year X, PC, MS. 2199; "Etat général des professeurs, ingénieurs, élèves et employés de l'Ecole nationale des ingénieurs géographes," year X, PC, MS. 2148; Margaret Bradley, "Financial Basis of Science in Paris 1790–1815," *Annals of Science* 36(1979):451–91.

51. Marcel Destombes, "De la chronique à l'histoire: Le globe terrestre monumental de Bergevin (1784–1795)," *Archives Internationales d'Histoire des Sciences* 27(1977):113–34.

52. Berthaut, *Ingénieurs géographes*, 1:126–61, is the best source for the national map archive.

53. Ibid., 145.

54. Broc, *Géographie des philosophes*, 466–74.

55. R. Jouanne, *Les origines du cadastre ornais* (Alençon: Imprimerie Alençonnaise, 1933), 26; Hugh D. Clout and Keith Sutton, "The 'Cadastre' as a Source for French Rural Studies," *Agricultural History* 43(1969):215–23; R. Herbin and A. Pebereau, *Le cadastre français* (Paris: Francis Lefebre, 1953).

56. Laprade's original project and Prony's report are in AN, F¹⁴2146.

57. Berthaut, *La carte de France*, 1:170 ff.

58. Ibid.

59. In a personal communication, Roger Hahn informed me of the existence of documents pertaining to Laplace's lifelong interest in geodesy that have not been the subject of scholarly study.

60. Gillispie, *Science and Polity*, 550.

61. Ibid., 549.

Chapter Three

1. Lloyd A. Brown, "The River in the Ocean," in *Essays Honoring Lawrence C. Wroth* (Portland, Me.: Athoensen Press, 1951), 69–84; quotation is from 71. I am indebted to Douglas Marshall for providing me with a copy of this essay. See also Louis de Vorsey, "Pioneer Charting of the Gulf Stream: The Contributions of Benjamin Franklin and William Gerard De Brahm," *Imago Mundi* 28(1976):105–20; idem, "The Gulf Stream on Eighteenth-Century Maps and Charts," *Map Collector*, no. 15 (June 1981):2–10.

2. Vorsey, "Pioneer Charting."

3. Brown, "River in the Ocean," 76–78.

4. Philip L. Richardson, "Benjamin Franklin and Timothy Folger's First Printed Chart of the Gulf Stream," *Science* 207 (8 February 1980):643–45.

5. Two copies are in Paris: BN, Cartes et Plans, Service Hydrographique, port. 117, nos. 7, 7¹; third copy is in Naval Library, Ministry of Defense, London.

6. John Noble Wilford, "Prints of Franklin's Chart of Gulf Stream Found," *New York Times*, 6 February 1980, A1, B7.

7. Benjamin Franklin, "A Letter from Dr. Franklin to Mr. Alphonsius Le Roy, member of several Academies at Paris, concerning secondary Maritime observations," August 1785, published in the *Transactions of the American Philosophical Society* 2(1786). It was accompanied by a chart of the North Atlantic, engraved by James Poupard, with an insert showing the Gulf Stream's current.

8. Today an infrared sensor aboard the Geostationary Operational Environmental Satellite of the National Oceanic and Atmospheric Administration senses the warmer waters of the Gulf Stream and converts that information into pictures and a map showing the current.

9. Charles Blagden, "On the Heat of the Water in the Gulf Stream," *Philosophical Transactions of the Royal Society* 7(1781):334.

10. Thomas Truxton, *Remarks . . . and Examples relating to the Latitude and Longitude* (Philadelphia, 1794), with "A General Chart of the World Showing the course of the Gulph [*sic*] Stream and various tracks to and from the East Indies . . ."; Jonathan Williams, "Memoir of Jonathan Williams on the use of the thermometer in discovering Banks, Soundings, etc.," *Transactions of the American Philosophical Society* 3(1793):82–100, which was accompanied by a chart by Williams showing four sailing tracks on the North Atlantic and a chart by Franklin with various

temperature readings for the same ocean; idem, "A Thermometrical Journal of the temperature of the atmosphere and Sea, on a voyage to and from Oporto, with explanatory observations thereon," *Transactions of the American Philosophical Society* 3(1793):194–202. This second piece was based upon a comparison of observations by Williams, Franklin, and Captain William Billings, who presented his own seventy-three-page journal to the American Philosophical Society in 1792. Williams's texts were published in book form as *Thermometrical Navigation* (Philadelphia, 1799). William Strickland made observations on a crossing to England in 1794 and wrote a report in 1798, "On the Use of the Thermometer in Navigation," *Transactions of the American Philosophical Society* 5(1802):90–103, which was accompanied by a chart of the Atlantic showing the Gulf Stream and a tributary of it toward Ireland and Scotland that Strickland believed to exist.

11. Williams, "Thermometrical Journal."

12. Arthur H. Robinson, "The Genealogy of the Isopleth," *Cartographic Journal* 8(1971):49–53; reprinted in *Surveying and Mapping* 32(1972):331–38, from which this quotation was taken (331–32).

13. Luigi Ferdinando Marsigli, *Histoire physique de la mer* (Amsterdam, 1725); Maris Longhena, *L'opera cartographica di L. F. Marsili,* Pubblicazioni dell'Instituto de Geografia, ser. A, no. 3 (Rome: University of Rome, 1933). See also Broc, *Géographie des philosophes,* 211–15; Margaret Deacon, *Scientists and the Sea, 1650–1900: A Study in Marine Science* (New York: Academic Press, 1971), viii–x, 175–76; Josef W. Konvitz, "Changing Concepts of the Sea, 1550–1950," *Terrae Incognitae* 11(1979):1–17.

14. Robinson, "Genealogy of the Isopleth," 332.

15. "Carte de partie du cours de la rivière de Garonne passant devant Bordeaux . . . ," 11 June 1729, Bibliothèque de l'Arsenal, 6439, fol. 157.

16. George Kish, "Early Thematic Mapping: The Work of Philippe Buache," *Imago Mundi* 28(1976):129–36.

17. This idea was clearly similar to Buffon's thinking on the formation of mountains, continents, and seas; Broc, *Géographie des philosophes,* 191.

18. MARS (1752), 399–416.

19. Kish, "Early Thematic Mapping," 131.

20. Numa Broc, "Un géographe dans son siècle: Philippe Buache (1700–1773)," *Dix-Huitième Siècle* 3(1971):222–35.

21. Kish, "Early Thematic Mapping," 132, and fig. 3, 133; Broc, *Géographie des philosophes,* 217–18.

22. Buache, "Explication du Globe Physique en relief Présenté au Roi le 6 Novembre 1757," BN, Cartes et Plans, Ge. FF. 13732–II.

23. Buache to Comte de Maurepas, January 1730, BN, Cartes et Plans, Ge. DD. 2334. "Réduction de la carte universelle de Mr. Halley sur le système de l'Amiant publiée en 1700; tout le plan géographique est comparé avec celui de Guill. Delisle par Philippe Buache son gendre (1732)," separate sheets of the Atlantic and Pacific, Bibliothèque de l'Institut, MS. 3804 (1), (2). For more information on isogonic maps, see John Cawood, "Terrestrial Magnetism and the Development of International Collaboration in the Early Nineteenth Century," *Annals of Science* 34(1977):551–87. In 1770, J. H. Lambert extended coverage of isogones to land surfaces.

24. Robinson, "Genealogy of the Isopleth," 332–33. François de Dainville, "De la profondeur à l'altitude: Des origines marines de l'expression cartographique du relief terrestre par côtes et courbes de niveau," in *Le navire et l'économie maritime du Moyen Age au dix-huitième siècle,* Travaux du 2ᵉ Colloque d'Histoire Maritime (Paris: Ecole Pratique des Hautes Etudes, 1958), 195–213; republished in *International Yearbook of Cartography* 2(1962):150–60; translated by Arthur H. Robinson as "From the Depths to the Heights," *Surveying and Mapping* 30(1970):389–403. See also Arthur H. Robinson, *Early Thematic Mapping in the History of Cartography* (Chicago: University of Chicago Press, 1982).

25. The Mountaine and Dodson essay on Halley's chart was published in 1757. I have seen a copy of it in BN, Cartes et Plans. The table is described in Taylor, *Haven-Finding Art,* 240.

26. A. H. W. Robinson, *Marine Cartography in Britain: A History of the Sea Chart to 1855* (Leicester: Leicester University Press, 1962); idem, "Marine Surveying in Britain during the Seventeenth

and Eighteenth Centuries," *Geographical Journal* 123(1957):449–56 idem, "The Charting of the Scottish Coasts," *Scottish Geographical Magazine* 74(1958):116–26; R. A. Skelton, "Captain James Cook as a Hydrographer," *Mariner's Mirror* 40(1952):92–119.

27. Robinson, "Marine Surveying."

28. Robinson, "Charting the Scottish Coasts," 121.

29. Methods for fixing a ship's position included resection by observation of three rays on shore objects (first described in English in 1764), the station pointer (invented about 1780), and marine triangulation based on the use of buoys. Of enormous importance also were the accurate lunar tables based on the work of Euler, Mayer, Clairaut, d'Alembert, Laplace, and Lagrange. Lunar tables were first used at sea by Lacaille on his return from the Cape of Good Hope in 1753–54. Howse, *Greenwich Time*, 60–67; Eric G. Forbes, "Mathematical Cosmography," in *The Ferment of Knowledge: Studies in the Historiography of Eighteenth-Century Science*, ed. Roy Porter and G. S. Rousseau, 417–48, (New York: Cambridge University Press, 1980). esp. 437–44, In *The Sky Explored: Celestial Cartography, 1500–1800* (New York: Alan R. Liss, 1979), A. J. Warner noted that some astronomers projected quadrants, compasses, and telescopes as new constellations in the heavens, thereby displacing religious and mythological symbols with more secular and scientific ones.

30. Alexander Dalrymple, *Essay on the Most Commodious Methods of Marine Surveying* (1771); Murdoch Mackenzie, Sr., *A Treatise of Maritim* [sic] *Surveying* (1774). See also Alun C. Davies, "The Life and Death of a Scientific Instrument: The Marine Chronometer, 1770–1920," *Annals of Science* 35(1978):509–25; and David S. Landes, *Revolution in Time: Clocks and the Making of the Modern World* (Cambridge: Belknap Press of Harvard University Press, 1983).

31. F. Russo, "L'hydrographie en France au XVII⁶ et XVIII⁶ siècles," in *Enseignement et diffusion des sciences en France au dix-huitième siècle*, ed. René Taton, 419–40 (Paris: Hermann, 1964); Frederick B. Artz, *The Development of Technical Education in France, 1500–1850* (Cleveland: Society for the History of Technology, 1966), 48–59, 102–9, 175 ff.

32. Gillispie, *Science and Polity*, 337–44.

33. P. J. Charliat, "L'Académie Royale de Marine et la révolution nautique au XVIII⁶ siècle," *Thalès* 3(1934):71–82; idem, "Le temps des grands voiliers," in Broc, *Géographie des philosophes*, 287–97.

34. Howse, *Greenwich Time*, 73–99; Landes, *Revolution in Time*, 162–70.

35. Harry Woolf, *The Transits of Venus* (Princeton: Princeton University Press, 1959). Delisle's map of 1760 is in BN, Cartes et Plans, Ge. D. 12821; his map of 1758 is Ge. D. 5077; Lalande's map of 1764 is Ge. DD. 2987–B(54). See also Laussedat's approach to mapping areas of the earth exposed to an eclipse in *Connaissance des temps* (1870), and in *Comptes-Rendues de l'Académie des Sciences* 70(1870):240 ff.

36. Broc, *Géographie des philosophes*, 283–84.

37. AN, Marine 3JJ, fol. 9, "Sur la correction des cartes de navigation."

38. Buache, "Sur le Cape Hinlopen," BN, Cartes et Plans, Ge. FF. 13732.

39. "Projet de carte de Méditerrannée," 1735, BN, Cartes et Plans, Service Hydrographique, port. 64, no. 12. A selection of these manuscript maps are in the Bibliothèque de l'Institut, MSS. 2721, 3804.

40. Josef W. Konvitz, "Alexander Dalrymple's Wind Scale for Mariners," *Mariner's Mirror* 69(1983):91–93.

41. AN, Marine 3JJ 4, fols. 1–5.

42. Buache to Comte de Maurepas, 2 February 1737, BN, Cartes et Plans, Ge. FF. 13732–II.

43. Charles-Pierre Claret de Fleurieu, *Observations sur la division hydrographique du Globe et changements proposés dans la nomenclature générale et particulière de l'hydrographie: Application du système métrique décimal à la hydrographie et aux calculs de la navigation* (Paris: Imprimerie de la République, year VIII). A more comprehensive survey of French hydrography would include the work of J. N. Bellin (1703–72), author of *L'hydrographie française* (1757) and *Le petit atlas maritime*, and J. B. d'Après de Mannevillette (1707–80), editor of *Le Neptune Oriental* (1745).

44. J. L. Dupain-Triel, *Recherches géographiques sur les hauteurs des plaines du royaume, sur les mers et leurs côtes presque pour tout le globe: Et sur les diverses espèces de montagnes* (Paris: Hérault, 1791), and accompanying map, "La France considérée dans les différentes hauteurs de ses plaines . . . ," BN, Cartes et Plans, Ge. D. 15126.

45. Dupain-Triel, "Carte de France, où L'on a essayé de donner la configuration de son territoire, par une nouvelle méthode de nivellements," BN, Cartes et Plans, Rés. Ge. DD. 7657.

46. J. L. Dupain-Triel, *Mémoire Explicatif de la Géographie perfectionnée par de nouvelles méthodes de nivellements, d'après Du Carla, méthodes spécialement avantageuses à la Navigation Intérieure, ainsi qu'aux Ponts et Chaussées; publiées pour la deuxième fois* (year XI [1804]).

Chapter Four

1. Martin J. S. Rudwick, "The Emergence of a Visual Language for Geological Science, 1760–1840," *History of Science* 14(1976)149–95.

2. Ibid., 177–82.

3. Numa Broc, *Les montagnes vues par les géographes et les naturalistes de langue française au XVIIIᵉ siècle* (Paris: Bibliothèque Nationale, 1969), 24. See also Walter Kirchner, "Mind, Mountain and History," *Journal of the History of Ideas* 11(1950):412–47; Marjorie Nicolson, *Mountain Gloom and Mountain Glory: The Development of the Aesthetics of the Infinite* (Ithaca, N.Y.: Cornell University Press, 1959); *Images de la montagne; De l'artiste à l'ordinateur, catalogue et essais* (Paris: Bibliothèque Nationale, 1984).

4. Louis Bourguet, *Lettres philosophiques sur la formation des sels et des crystaux, avec un Mémoire sur la théorie de la terre* (Amsterdam: François l'Honoré, 1729), esp. 177–90.

5. Broc, *Montagnes*, 118.

6. Rhoda Rapoport, "The Early Disputes between Lavoisier and Monnet, 1777–1781," *British Journal for the History of Science* 4(1969):233–44; idem, "The Geological Atlas of Guettard, Lavoisier and Monnet: Conflicting Views of the Nature of Geology," in *Toward a History of Geology*, ed. Cecil J. Schneer, 272–87 (Cambridge: MIT Press, 1969); idem, "Lavoisier's Geological Activities, 1763–1792," *Isis* 58(1967); 375–84; idem, "Problems and Sources in the History of Geology, 1794–1810," *History of Science* 3(1964):60–77. On Guettard, see André Cailleux, "The Geological Map of North America (1752)," in *Two Hundred Years of Geology in America*, ed. Cecil J. Schneer, 43–52 (Hanover, N.H.: University Press of New England, 1979). On Buache and Guettard, see Kish, "Early Thematic Maps."

7. Rudwick, "Emergence of a Visual Language," 161.

8. Rapoport, "Geological Atlas," 277.

9. Broc, *Montagnes*, 69.

10. Buache, "Carte générale de Languedoc subdivisée par Terreins de Fleuves et de leurs Rivières. Le tout indiqué par la suite des Chaines de Montagnes qui traversent les XXII diocèses de cette province. Dressée suivant le système présentée en 1752 à l'Académie des sciences par Phil. Buache," BN, Cartes et Plans, Ge. B. 2384.

11. Regardless of the limitations of theory, scientists relied on Buache's maps until better theories as well as better maps came along.

12. Rapoport, "Geological Atlas," 283.

13. Ibid., 287.

14. Gillispie, *Science and Polity*, 172; Kenneth L. Taylor, "Nicolas Demarest and Geology in the Eighteenth Century," in *Toward a History of Geology*, ed. Schneer, 339–56.

15. Taylor, "Demarest and Geology," 340–41.

16. Ibid., 346, n. 23.

17. Rudwick, "Emergence of a Visual Language," 163–64.

18. Gillispie, *Science and Polity*, 382–83.

19. Henri Pigonneau and Alfred de Foville, *L'administration de l'agriculture au contrôle-générale des finances (1785–7): Procès-verbaux et rapports* (Paris: Guillaumin, 1882), esp. 68–88, 114.

20. On British geology, see Roy Porter, *The Making of Geology: Earth Science in Britain, 1660–1815* (Cambridge: Cambridge University Press, 1978); Rudwick, "The Emergence of a Visual Language"; Joan Eyles, "William Smith: Some Aspects of His Life and Work," in *Toward a History of Geology*, ed. Schneer, 142–58; R. C. Boud, "The Early Development of British Geological Maps," *Imago Mundi* 27(1975):73–96; idem, "Aaron Arrowsmith's Topographical Map of Scotland and John Macculloch's Geological Survey," *Canadian Cartographer* (now *Cartographica*) 11(1974):24–34; idem, "Samuel Hibbert and the Early Geological Mapping of the Shetland Islands," *Cartographic Journal* 14(1977):81–88; and John G. C. M. Fuller, "The Industrial Basis of Stratigraphy: John Strachey, 1671–1743, and William Smith, 1769–1839," *American Association of Petroleum Geologists, Bulletin* 53(1969):2256–73.

21. Rudwick, "Emergence of a Visual Language," 170.

22. Broc, *Montagnes*, 68.

23. John Wolter, "The Heights of Mountains and the Lengths of Rivers," Library of Congress, *Quarterly Journal* 29(1972), reprinted in *Surveying and Mapping* 32(1972):313–29.

24. Broc, *Montagnes* 176, 187.

25. François Pasumot, *Voyages physiques dans les Pyrénées en 1788 et 1789: Histoire naturelle d'une partie de ces montagnes, particulièrement des environs de Barège, Bagnères, Cautères et Gavarnie. Avec des cartes géographiques* (Paris: L'Imprimerie de Le Cleres, 1797).

26. Jean-Marcel Cadet, "Carte minéralogique de l'isle de Corse," BN, Cartes et Plans, Ge. B. 8199; the map was published in Cadet's *Vérités physiques élémentaires pour l'étude de l'histoire naturelle prouvés par l'état du sol de la Corse* (1789).

27. Giraud-Soulavie, *Géographie de la nature, ou Distribution naturelle des trois règnes sur la surface de la terre, suivie de la carte minéralogique, botanique, etc. du Vivarais ou cette distribution naturelle est représentée. Ouvrage qui sert de préliminaire à l'histoire naturelle de la France meridionale, etc., dont on va publier les deux premiers volumes, etc., à l'histoire ancienne et physique du Globe terrestre*, vol. 1 (Paris: Dupain-Triel, 1780), vol. 2 (1788): idem, *Prospectus de l'histoire naturelle de la France meridionale* (Nîmes: Castor Belle, 1780); "Carte géographique de la nature, ou Disposition naturelle des minéraux, végétaux etc. observé en Vivarais. Dressée par le sr Dupain-Triel . . . d'après les ouvrages de Mr. l'Abbé Giraud-Soulavie" (1780), opposite p. 16 of *Géographie*, vol. 1.

28. L'Abbé Palassou, *Essai sur la minéralogique des Monts-Pyrénées, suivi d'un catalogue des plantes observées dans cette chaine de montagnes* (Paris: Didot, 1781).

29. Broc, *Montagnes*, p. 28 and figs. 6, 7.

30. Ibid., 80, 84, 87, 91 for tables comparing several values for altitude of mountains as measured in the eighteenth century with their true value.

31. M. Reboul, "Nivellement des principaux sommets de la Chaîne des Pyrénées," *Annales de Chimie et de Physique* 5(1817):234–60.

32. Close, *Early Years of the Ordnance Survey*, 7.

33. Ibid., 142. See also Yolande Jones, "Relief Portrayal on 19th Century British Military Maps," *Cartographic Journal* 11(1974):19–32.

34. Berthaut, *Ingénieurs géographes*, vol. 1; Blanchard, *Ingénieurs du "roy" de Louis XIV à Louis XVI*.

35. On the Savoy cadaster, see *Le cadastre sarde de 1730 en Savoie*.

36. Marcel Huguenin, "La cartographie des Alpes françaises avant Cassini," *Bulletin d'Information de l'Association des Ingénieurs Géographes*, no. 10 (March 1958):89–106; no. 12 (November 1958):105–26.

37. Instructions in 1761 about field surveys to engineers stated: "Figurer et exprimer de même, avec intelligence, par le plus ou moins d'inclinaison des hachures du dessin, le degré d'élévation des montagnes. Distinguer si elles sont accessibles ou inaccessibles . . . , ainsi que le commandement des dites montagnes les unes sur les autres. . . . Il faut aussi exprimer les pentes et la nature des chemins qui montent, descendent et tournent les montagnes. Il faudra donner pour cet effet des coups de niveau [cross sections] pour s'assurer de la pente qu'il peut y avoir par toise, afin de savoir par ce moyen la quantité de chevaux pour y faire passer des voitures plus ou moins chargées, ou de l'artillerie de plus ou moins gros calibre." Berthaut, *Ingénieurs géographes*, 1:28. Instructions in 1772 added that because the degree of slope is rendered too impressionistically by hatch lines, engineers were to indicate accessibility by adding a letter—P for a man on foot, S for man on horseback, T for wheeled vehicles. Ibid., 47. See also Douglas W. Marshall, "Instructions for a Military Survey in 1779," *Cartographica* 18(1981):1–12.

38. L. N. Lespinasse, *Traité du lavis des plans appliqué principalement aux reconnaissances militaires*, 2d ed. (Paris: Maginel, Anselin and Pochard, 1818), 6.

39. On the conflict between military engineers and the staff of the Cassini map: d'Arçon to M. de Vault, letter of 18 August 1776, SHAT, correspondence topographique, A.1; quoted also by Berthaut, *Ingénieurs géographes*, 1:72–74. Military engineers were occasionally sent into mountainous regions of other countries (e.g., Switzerland) on espionage missions; Ibid., 59–62.

40. On relief maps after 1789, see Berthaut, *Ingénieurs géographes*, 2:296–304. See also the note by George A. Rothrock, "Musée des Plans-Reliefs," *French Historical Studies* 6(1969):253–56, and the

small book, well illustrated, by Catherine Brisac, *Le Musée des Plans-Reliefs: Hôtel National des Invalides* (Paris: Pygmalion/Gérard Watelet, 1981).

41. Skelton, "Cartography," 612.

42. Louis Milet de Mureau, "Mémoire pour faciliter les moyens de projetter dans les pays de montagne avec le seul secour du plan du terrain levé exactement" (1749), Archives du Génie (Vincennes), art. 21, sect. 1, para. 11, carton 4, no. 5.

43. Gillispie, *Science and Polity*, 506.

44. Ibid., 533.

45. Jean-Baptiste Meusnier, "Mémoire sur la détermination du site" (1777), Archives du Génie (Vincennes), art. 21, sect. 1, para. 1, carton 6, no. 12.

46. On Meusnier's work at Cherbourg, see Bibliothèque du Génie (Paris), MS. in-fol. 131, manuscript register and correspondence; "Notices historiques sur les travaux de la rade de Cherbourg, par l'ingénieur Meusnier" (1791), PC, MS. 1000; Jean-Baptiste Meusnier, "Instruction pour terminer le travail des sondes de la rade de Cherbourg" (1790), Archives du Génie (Vincennes), art. 8, sect. 1, carton 5, no. 25; and the manuscript map of the harbor, IGN, no. 262.

47. *Mémorial du Dépôt Général de la Guerre* (Paris: Picquet, 1829), vol. 1 and 2:1–140.

48. Pierre-Simon Girard, *Recherches sur les eaux publiques de Paris, les distributions successives qui en ont été faites, et les divers projets qui ont été proposés pour en augmenter le volume* (Paris: Imprimerie Nationale, 1812). Emile Levasseur suggested to Haussmann that contour lines be added for elevation at one-meter intervals on the new geodetic map of Paris Haussmann had made, but Haussmann did not agree that this was necessary. He did accept a proposal by Levasseur to place bronze disks in the pavement at points where triangulation scaffolds were erected, since in principle triangulation should be based on permanent, verifiable points of reference. On maps of Paris between 1660 and 1820, see Josef W. Konvitz, *The Urban Millennium: The City-Building Process from the Early Middle Ages to the Present* (Carbondale: Southern Illinois University Press, 1985), 78–95.

49. S. F. Lacroix, *Introduction à la géographie mathématique et critique et à la géographie physique* (Paris: Dentu, 1811).

Chapter Five

1. Robinson, *Early Thematic Mapping*, 16–17.

2. Naudin, "Inventaire des cartes, plans et mémoires apartenant au Roy pour l'usage du Ministère de la Guerre . . . ," Bibliothèque du Génie (Paris), MS. in-fol. 209. On maps for the Canal des Deux-Mers, see François de Dainville, *Les cartes anciennes de Languedoc XVIe–XVIIIe siècles* (Montpellier: Société Languedocienne de Géographie, 1961), 45.

3. BN, Cartes et Plans, Rés. Ge. AA. 2053.

4. "Mémoire sur plusieurs canaux projettés en Lorraine" (1751), BN, Cartes et Plans, Ge. DD. 559. The 1756 map is "Esquisse ou plan général des jonctions de rivières projettées en Lorraine . . . dont les trois projets n'ont jamais été réunis sous un seul point de vue . . . ," AN, N III Meuse 1, scale of 1:177,000.

5. De Kermadec du Moustier, *Projet d'une description géographique oeconomique et historique de la province de Bretagne, imprimé par ordre des Etats de Bretagne, tenus à Rennes au mois de decembre 1746. Par un membre de l'Assemblée de l'ordre de noblesse* (Rennes, 1748); *Précis des opérations relatives à la navigation intérieure de Bretagne* (Rennes: Vatar, pour les Etats de Bretagne, 1785); *Rapport de M. L'Abbé Bossut, Rochon, Fourcroy, Condorcet . . . sur la navigation intérieure de la Bretagne* (Paris: Imprimerie Royale, 1786); Gillmor, *Coulomb*, 52–60. The maps made in 1784 are in the Archives Départementales, Ille et Vilaine (Rennes), but I have not examined them. Maps in Paris related to the schemes of 1783–86 are Gotrot (ingénieur géographe), "Carte topographique des passes, rades et ports de Saint-Malo" (1783), AN, N I Ille et Vilaine 2; Picart de Norcy, "Carte générale des fleuves des rivières et des ruisseaux de Bretagne, pour servir à la navigation intérieure de cette province" (1784), BN, Cartes et Plans, Ge. C. 3577.

6. Jean Delagrive, "Cours de la Seine et des rivières qui y affluent levé sur les lieux par ordre de Mr. Le Président Turgot . . . ," (1738), BN, Cartes et Plans, Rés. Ge. CC. 1389, and "Recueil des rivières qui se rendent dans le Fleuve de la Seine . . . ," BN, Cartes et Plans, Ge. DD. 5533.

7. "Mémoire sur les travaux géographiques de la ville; mémoire concernant le receuil géographique du cours de la Seine, exécuté en 1767 par Phil. Buache," BN, Cartes et Plans, Ge. DD. 2334. See also Broc, "Un géographe dans son siècle: Philippe Buache."

8. "Carte générale du cours de la Seine de Paris à Rouen" (October–November 1766), AN, F¹⁴10078¹.

9. BN, Cartes et Plans, Ge. AA. 1374. See also a map made in 1817, "Carte des rivières et ruisseaux du bassin de la Seine servant à l'approvisionnement de Paris divisé en départements avec l'indication des flottages en train et à bois perdu, des pertuis, écluses, vannes, portes marinières etc. Pour l'ouvrage intitulé Code du Commerce des Bois det Charbons de Bois par M. Dupin, Avocat," BN, Cartes et Plans, Ge. D. 5547.

10. Buache, who developed the idea of a schematic outline of the Seine and its tributaries in 1752, made a model table on which to record information about where tributaries entered the Seine. Buache also made twenty-seven manuscript maps of the Seine's tributaries, annotated with notes on obstacles to navigation, areas prone to flooding, woods and forests, etc., BN, Cartes et Plans, Ge. DD. 5529. Buache included a bar graph showing elevation of the Seine in Paris in "Exposé de divers objets de la géographie physique concernant les bassins terrestres des fleuves et rivières qui arrosent la France, dont on donne quelques détails et en particulier celui de la Seine," MARS (1767, pub. 1770), 504–9. This work should be understood as a continuation of Buache's earlier work of 1753 on rivers and basins as part of his proposal for a relief globe.

11. M. de Montgéry, *Mémoire sur les moyens de rendre Paris port de Mer*, extrait des *Annales de l'industrie nationale et étrangère* (Paris: Bachelier, 1824); Pierre Forfait, "Expériences faites par ordre du gouvernement sur la navigation de la Seine," *Mémoires de l'Institut, Sciences Mathématiques et Physiques* 1 (year VI):120–68. A survey of the Seine in 1829–30 produced a map showing areas of fishing, mills, towpaths, and stretchs difficult to navigate (shaded gray). Julien Coïc, *Reconnaissances de la Seine de Rouen à Saint-Denis en 1829 et 1830 et travaux proposés pour rendre cette partie de la Seine facilement navigable* (Paris: A. Barbier, 1830).

12. Konvitz, *Urban Millennium*.

13. The Ponts et Chaussées was not in control of roads in the pays d'élection, only in the pays d'état. On the Ponts et Chaussées, see Vignon, *Etudes historiques sur l'administration des voies publiques*, and Petot, *Histoire de l'administration des Ponts et Chaussées*. On Coulomb's memorandum of 1777 about the organization and roles of fortifications engineers, see Gillmor, *Coulomb*, 29–32, 255–61.

14. Arbellot, "Grande mutation," 778–79. See Perronet, "Instruction pour les plans des principales routes et chemins de la province d'Artois qui doivent être levées pour le Roy" (18 March 1762), PC, MS. 2254.

15. Arbellot, "Grande mutation," 778.

16. Ibid., 786–91.

17. Perronet, "Instruction pour les plans."

18. Arbellot, "Grande mutation."

19. Ibid., 790.

20. This document is in PC, MS. "Ancien côte" XII.3.

21. "Instruction sur la carte itinéraire que chaque ingénieur en chef doit fourner de son Département," with an attached "Carte gravée et réduite au quart de l'échelle de la Carte générale de France" (1 July 1792), AN, F¹⁴954². The "Carte hydrographique de la République française" was to be eight feet square. "Rapport sur la carte hydrographique de la République française," PC, MS. 2402.

22. André Bourde, *Agronomie et agronomes en France au XVIII⁰ siècle* (Paris: SEVPEN, 1967).

23. Henri de Goyon de la Plombaine, *La France agricole et marchande* (Avignon, 1762).

24. Papiers Condorcet, Bibliothèque de l'Institut, MS. 866, fol. 8. Condorcet's comments were in reference to a letter addressed by Cornuau to the controller general, dated 11 November 1774.

25. Comment appeared in text entitled "Dépenses faites en 1777 en Dauphiné pour la Carte de Dauphiné et de Provence," SHAT A¹3703, fol. 112.

26. Ibid., fol. 62.

27. De Dainville, *Cartes anciennes de Languedoc*, 187.

28. Jean-Antoine Fabre, "Essai sur la théorie des torrens et rivières des pays de montagnes," presented to the Academy 2 December 1780; idem, *Traité complet sur la théorie et la pratique du nivellement* (Draguignan: Chez Fabre, [probably 1790s]).

29. Etienne Claude Baron de Marivetz, "Discours préliminaire et prospectus d'un traité général de géographie physique et particulièrement de celle du Royaume de France . . ." (Paris: De Guillau, 1779).

30. Anon. [de Marivetz], *Observations sur quelques objets d'utilité publique, pour servir de Prospectus à la seconde partie de la Physique du Monde, ou A la Carte Hydrographique de la France, et au Traité général de la Navigation intérieure de ce monde* (Paris: Visse, 1786), 44.

31. Ibid., 73, for a description of the maps of France; 144–52, for the work of the Ponts et Chaussées; 281–82, for an appeal to the state.

32. Paris, 1788.

33. Emiland-Marie Gauthey, "Carte des chaines de montagnes de la France, de ses principales rivières et des principaux canaux de navigation, faits ou à faire, dans ce royaume . . ." (1782), BN, Cartes et Plans, Ge. D. 14335; idem, "Carte de la France par bassins, avec profils" (no date), BN, Cartes et Plans, Ge. DD. 160; Nicolas Fer de la Nouerrre, "Carte élémentaire de la navigation du royaume," presented to the Academy in 1787, BN, Cartes et Plans, Ge. C. 1269.

34. Jean-Louis Dupain-Triel, "Carte générale des fleuves, des rivières et des principaux ruisseaux de la France avec les canaux existants ou même projettés à l'usage de la navigation intérieure du Royaume dédiée à Messieurs les Intendants du Commerce" (1781), BN, Cartes et Plans, Ge. C. 9878.

35. Jean-Louis Dupain-Triel, "Tableau géographique de la navigation intérieure du territoire républicain français, offrant le cours soit de ses fleuves, rivières et ruisseaux, soit de ses canaux, tant exécutés que projettés dans ses 86 départements," accompanied by a table, "Etat actuel et général de la navigation intérieur de la France" (1793), BN, Cartes et Plans, Ge. FF. 10963.

36. Jean-Louis Dupain-Triel, *Essai sur les moyens d'arriver à une hydrographie complète de l'intérieur de la République* (Paris: Impimerie Stoupe, 1796), PC 4897/C276.

37. Sanson, "Supplément à l'instruction sur le service des ingénieurs géographes en campagne" (year X), SHAT, Mémoires reconnaissances, 1121.

38. Petot, *Histoire de l'administration des Ponts et Chaussées*, 402. See also Pierre Pinon and Annie Kriegel, "L'achèvement des canaux sous la Restauration et la Monarchie de Juillet," *Annales des Ponts et Chaussées*, n.s., no. 19(1981):72–83.

39. Pierre-Louis Dupuis-Torcy and Mathurin-Jacques Brisson, "Essai sur l'art de projeter les canaux de navigation," *Journal de l'Ecole Polytechnique* 7(1808):262–88.

40. Keith Baker, *Condorcet: From Natural Philosophy to Social Mathematics* (Chicago: University of Chicago Press, 1975), 67.

41. Ibid., 69.

42. F. M. L. Thompson, *Chartered Surveyors: The Growth of a Profession* (London: Routledge and Kegan Paul, 1968), 55.

Chapter Six

1. Robinson, *Thematic Mapping*, 17.

2. Ibid., 16. I have used Robinson as a source, but wherever possible I have also consulted original documents. Robinson, for example, quoted Quételet's published reference to a letter he received from Guerry. I located that letter in the Quételet papers in the Académie Royale de Belgique, and I also raise a question Robinson did not ask about Quételet's lack of interest in thematic mapping after 1835. Our purposes are different. By including more material on French weather maps and on medical maps, for example, I do not invalidate or contradict what Robinson has written on related matters.

3. "La France commerçante," BN, Cartes et Plans, Ge. D. 6818.

4. Marie Le Masson Le Golft, "Esquisse d'un tableau général du genre humain, ou l'on s'apperçoit d'un seul coup d'oeil les religions et les moeurs des différents peuples, les climats sous lesquels ils habitent et les principales variétés de forme et de couleur de chacun d'eux," BN, Cartes et Plans, Ge. C. 8674.

5. Untitled, anonymous administrative and economic map of France, BN, Cartes et Plans, Rés. Ge. A. 1106. See also Myriem Foncin, "A Manuscript Economic Map of France (End of the XVIIIth Century)," *Imago Mundi* 19(1965):51–55.

6. Robinson, *Thematic Mapping*, 43.

7. For a discussion of the origin of the concept of the rate of change, see John U. Nef, *Cultural Foundations of Industrial Civilization* (Cambridge: Cambridge University Press, 1958), 12 ff.

8. On Playfair, see H. G. Funkhauser, and H. M. Walker, "Playfair and His Charts," *Economic History* 3(1935): 103–9. More generally, see H. G. Funkhauser, "Historical Development of the Graphical Representation of Statistical Data," *Osiris* 3(1937):269–404. Both have good bibliographies of Playfair's publications. The quotation is from Funkhauser, "Historical Development," 289.

9. Funkhauser, "Historical Development," 290.

10. Robinson, *Thematic Mapping*, 206; on 141, Robinson mentions an economic map of Germany and Austria from 1796, with 126 symbols, which is almost illegible.

11. Funkhauser, "Historical Development," 292.

12. Ibid., 289.

13. William Playfair, *An Inquiry into the Permanent Causes of the Decline and Fall of Powerful and Wealthy Nations . . . Designed to show how the Prosperity of the British Empire may be prolonged* (London: Marchant, for Greenland and Norris, 1805), 214.

14. Ibid., legend for chart 1.

15. Jean-Claude Perrot, *L'age d'or de la statistique régionale française (an IV–1804)* (Paris: Société des Etudes Robespierristes, 1977).

16. P. E. Herbin and J. Peuchet, *Statistique générale et particulière de la France et de ses colonies avec une nouvelle description topographique, physique, agricole, politique, industrielle et commmerciale de cet état* (Paris: Buisson, 1803), 1:vii, xxvii, xxviii.

17. Funkhauser, "Historical Development," 308–9.

18. J. L. Wolff, "Essai de carte géologique et synoptique du Département de l'Ourte et de ses environs," engraved by L. Jehotte, Liège, BN, Cartes et Plans, Ge. D. 22101.

19. M. le Baron Coquebert de Montbret, "Essai d'une carte agricole de la France, des Pays-Bas et de quelques contrées voisines" (38.0 cm × 37.5 cm; 1:3,700,000), BN, Cartes et Plans, Sg. D. 187. The color code is white = regions without vineyards; pink = vineyards; green = olive trees; orange = orange trees.

20. Coquebert de Montbret's speech is the archives of the Academy of Sciences, "pochette de séance" for 19 February 1821. Relevant publications by Omalius d'Halloy are "Essai sur la géologie du Nord de la France," *Journal des Mines*, no. 140 (1808), reprinted by Bossange et Masson in Paris, 1809; "Observations sur un essai de carte géologique des Pays-Bas, de la France et de quelques contrées voisines," *Annales des Mines* (1822); and *Mémoires pour servir à la Description Géologique des Pays-Bas, de la France et de quelques contrées voisines* (Namur: D. Gerard, 1828). Their letters are in the "correspondence scientifique" of Omalius d'Halloy collected by the Académie Royale des Sciences, Brussels. Of special interest are letters by Coquebert de Montbret dated 26 November 1821 and 3 November 1822.

21. Broc, *Géographie des philosophes*, 464; Perrot, *Age d'or de la statistique régionale*, 65; the quotation is from Frank E. Melvin, *Napoleon's Navigational System: A Study of Trade Control during the Continental Blockade* (New York: Appleton, 1919), 371.

22. Omalius d'Halloy, "Esquisse d'une carte géologique de la France, des Pays-Bas et de quelques contrées voisines," BN, Cartes et Plans, Ge. D. 16778.

23. Robinson, *Thematic Mapping*, 102 and fig. 44 on 103.

24. Ibid.

25. Richard Hartshorne, "The Concept of Geography as a Science of Space, from Kant and Humboldt to Hettner," American Association of Geographers, *Annals* 48(1958):97–108; Edmunds V. Bunkse, "Humboldt and an Aesthetic Tradition in Geography," *Geographical Review* 71(1981):127–46.

26. Cawood, "Terrestrial Magnetism."

27. Robinson, *Thematic Mapping*, 56, 216; Arthur H. Robinson and Helen Wallis, "Humboldt's Map of Isothermal Lines: A Milestone in Thematic Cartography," *Cartographic Journal* 4(1967):19–23.

28. Robinson, *Thematic Mapping*, 96–99.

29. Petot, *Histoire des Ponts et Chaussées*, 373 ff.; Gillispie, *Science and Polity*, 482–87, 492.

30. "Instruction de 1775," arts. 20–22.

31. Brooke Hindle, *Emulation and Invention* (New York: New York University Press, 1981), esp. 22, 48, 79.

32. Eugene S. Ferguson, "The Mind's Eye: Nonverbal Thought in Technology," *Science* 197(1977):827–36; quotations from 835 and 828 in that order.

33. I have written an article for FMR about maps that display images of imaginary territories by students at the Ecole Royale des Ponts et Chaussées, to be published in 1986. Its provisional title is "Recreating France: The Cartographic Vision of Civil Engineers at the End of the Old Regime."

34. David Billington, *The Tower and the Bridge* (New York: Basic Books, 1983), emphasizes the aesthetic dimension of engineering design.

35. Petot, *Histoire des Ponts et Chaussées*, 372–75.

36. L. Pearce Williams, "Science, Education and the French Revolution," *Isis* 44(1953):311–30; Janis Langin, "Sur la première organisation de l'Ecole Polytechnique, texte de l'arrêté du 6 frimaire an III," *Revue de l'Histoire des Sciences* 33(1980):289–313.

37. P. J. Booker, "Gaspard Monge (1746–1818) and His Effect on Engineering Drawing and Technical Education," Newcomen Society, *Transactions* 34(1961–62):15–36; Alberto Pérez-Gomez, *Architecture and the Crisis of Modern Science* (Cambridge: MIT Press, 1983), 279–95. For detailed information on cartography in the curriculum, I consulted the published "Programmes de l'Enseignement de l'Ecole Impériale Polytechnique arrêtés par le conseil de perfectionnement" and the minutes of the council's meetings preserved in the archives of the Ecole Polytechnique.

38. "Travail de la commission chargée en l'an XI, de choisir, classer et compléter les ouvrages nécessaires à l'instruction des Elèves de l'Artillerie et du Génie" (7 April 1803), Archives du Génie (Vincennes), art. 18, sect. 2, carton 1, no. 2.

39. Artz, *Development of Technical Education in France*, 165.

40. Ibid.; Berthaut, *Ingénieurs géographes*, 2:148–49; Ecole Polytechnique, *Livre du centenaire, 1794–1894*, 3 vols. (Paris: Gauthiers-Villars, 1894), vol. 2 (services militaires), 279–80.

41. Archives de l'Ecole Polytechnique, Conseil de Perfectionnement, 1816–1830, vol. 5, Programmes des arts graphiques pour 1819–1820, séance du 30 décembre 1819.

42. Terry Shinn, *Savoir scientifique et pouvoir sociale: L'Ecole Polytechnique, 1794–1914* (Paris: Presses de la Fondation Nationale des Sciences Politiques, 1980), 29.

43. Lloyd G. Stevenson, "Putting Disease on the Map: The Early Use of Spot Maps in the Study of Yellow Fever," *Journal of the History of Medicine and Allied Sciences* 20(1965):227–61; Jean Meyer, ed., *Médecins, climat et épidémies à la fin du XVIIIᵉ siècle* (Paris and The Hague: Mouton, 1972).

44. Archives de la Société Royale de Médecine, "Arrêt du conseil d'état," 24 April 1786, MS. 114, dossier 16; Martin-Solon, Mestivier, Villerme, Thillaye, and Villeneuve, "Rapport général sur les épidémies qui ont regné en France depuis 1771 jusqu'à 1830 exclusivement, et dont les relations sont parvenus à l'Académie," *Mémoires de l'Académie Royale de Médecine* 3(1833):377–429.

45. E. W. Gilbert, "Pioneer Maps of Health and Disease in England," *Geographical Journal* 124(1958): 172–83. Robinson, *Thematic Mapping*, 170–74; Saul Jarcho, "Yellow Fever, Cholera and the Beginnings of Medical Cartography," *Journal of the History of Medicine and Allied Sciences* 25(1970):131–42.

46. Minutes of the Académie Royale de Médecine, meeting of 17 June 1833.

47. M. Hellis, *Souvenirs du choléra en 1832* (Paris: Ballière, Delaunay, 1833).

48. Robinson, *Thematic Mapping*, 174–75; J. Fr. Malagaigne, "Recherches sur la fréquence des hernies selon les sexes, les âges et relativement à la population," *Annales d'Hygiène Publique et de Médecine Légale* 24(1980):1–54.

49. J. Ch. M. Boudin, *Traité de géographie et de statistiques médicales*, 2 vols. (Paris: Ballière, 1857).

50. On earlier maps of weather conditions by Halley and Humboldt and on nineteenth-century atlases, see Robinson, *Thematic Mapping*, 69–79.

51. Donald R. Whitnah, *A History of the United States Weather Bureau* (Urbana: University of Illinois Press, 1961).

52. Correspondence of Adolphe Quételet, Bache to Quételet, letter dated 29 June 1843, Académie Royale de Belgique, dossier 271.

53. Adolphe Quételet, *Sur le climat de la Belgique* (Brussels: Hayez: 1849), 76–78, 89.

54. H. E. L. Mellersch, *Fitzroy of the Beagle* (London: Rupert Hart-Davis, 1968), 263–81.

55. Francis Galton, *Meteorographica, or Methods of Mapping the Weather, illustrated by upwards of 600 printed and lithographed diagrams referring to the weather of a large part of Europe during the month of December 1861* (London and Cambridge: Macmillan, 1863).

56. "Carte relative à l'orage du 13 juillet 1788" (96.5 cm × 170 cm; 1:350,000), AN, NN 51.

57. A. Danjon, "Le Verrier, créateur de la météorologie," *La Météorologie*, ser. 4, no. 1(1946):863–82.

58. Robinson, *Thematic Mapping*, 156–57.

59. Ibid., 157.

60. Ibid., 199.

61. Roger Chartier, "Les deux Frances: Histoire d'une géographie," *Cahiers d'Histoire* 23(1978):393–415.

62. Robinson, *Thematic Mapping*, 112–13.

63. Ibid., 200–01.

64. Ibid., 158–59. Funkhauser wrote that Guerry "would certainly have attained greater prominence in the field of moral statistics if he had not happened to have lived during the time of a Quetelet." "Historical Development," 304.

65. Correspondence of Adolphe Quételet, letter of Guerry to Quételet, 4 August 1833, Académie Royale de Belgique, dossier 1213.

66. George Sarton, "Preface to Volume XXIII of Isis (Quetelet)," *Isis* 23(1935):5–24, esp. p. 9.

67. Funkhauser, "Historical Development," 310.

68. Adolphe d'Angeville, *Essai sur la statistique de la population française considérée sous quelques-uns de ses rapports physiques et moraux* (Bourg: Dufour, 1836; reprinted with an introduction by Emmanuel Le Roy Ladurie, "Un théoricien du développement," v–xxxvi, Paris and The Hague: Mouton, 1969).

69. Ibid., xiv.

70. Ibid., 15–16.

71. Arthur H. Robinson, "The 1837 Maps of Henry Drury Harness," *Geographical Journal* 121(1955):440–50.

72. Robinson, *Thematic Mapping*, 147; see also 118, 208–9.

73. Ibid., 149. Alphonse Belpaire, *Notice sur les cartes du mouvement des transports en Belgique* (Brussels: Vandermalen, 1847).

74. A. de Humboldt, "Sur les lignes isothermes," *Annales de Chimie et de Physique* 4(1817):102–11; Robinson, *Thematic Mapping*, 70–72.

75. Robinson, *Thematic Mapping*.

76. Robinson, "Genealogy of the Isopleth," 49–50.

77. Jean Elie de Beaumont, Gabriel Lamé, and Augustin-Louis Cauchy, "Rapport sur un Mémoire de M. Léon Lalanne, qui a pour objet la substitution de plans topographiques à des tables numériques à double entrée," *Comptes-Rendus de l'Académie des Sciences* 17(1843):492–94.

78. Funkhauser, "Historical Development," 370–71.

79. Morlet, "Recherches sur les centres de figure," *Comptes-Rendus de l'Académie des Sciences* 20(1845):300–303.

80. Léon Lalanne, "Remarques à l'occasion du Mémoire de M. Morlet sur les centres de figures, et réflexions sur la représentation graphique de divers éléments relatifs à la population," *Comptes-Rendus de l'Académie des Sciences* 20(1845):438–41; quotation is from 440; my translation.

81. Lalanne's most complete statement on the subject of different kinds of contour lines appeared in a publication for civil engineers: "Sur les tables graphiques et sur la géométrie anamorphique appliquée à divers questions qui se rattachent à l'art de l'ingénieur," *Annales des Ponts et Chaussées* 11(1846): "Mémoires et documents," 1–69.

82. Funkhauser, "Historical Development," 302; Robinson, "Geneology of the Isopleth," 52.

83. *Rapport au Roi par le conseil de perfectionnement de l'Ecole royale polytechnique, sessions de 1817–1818* (Paris: Imprimerie Royale, 1818), 11–12.

84. On Minard, see François de Dainville, "Les bases d'une cartographie industrielle de l'Europe au XIXᵉ siècle," in *L'industrialisation en Europe au dix-neuvième siècle: Cartographie et typologie,*

15–33, Colloques Internationaux du CNRS, Lyon, 1970 (Paris: Bibliothèque Nationale, 1972); Arthur H. Robinson, "The Thematic Maps of Charles Joseph Minard," *Imago Mundi* 21(1967):95–108.

85. Charles Joseph Minard, *Projet de canal et de chemin de fer pour le transport de pavés à Paris* (Paris: Delaforest, 1826).

86. Robinson, "Maps of Minard," 98.

87. Charles Joseph Minard, "Notions élémentaires d'économie politique appliquées aux travaux publics," *Annales des Ponts et Chaussées* 19(1850).

88. Charles Joseph Minard, Des *tableaux graphiques et des cartes figuratives* (Paris: Thunot, 1862).

89. See Robinson, "Maps of Minard," 106–8 for a definitive cartobibliography.

90. De Dainville, "Bases d'une cartographie industrielle," 32–33; Funkhauser, "Historical Development," 331, 333.

Bibliography

Only materials published after 1850 are listed here; other documents are only cited in the text. A complete list of all the documents I consulted in the course of my research is impossible to provide here, since it would include literally hundreds of maps. I am, however, ready to correspond with any interested reader about primary sources.

Albitreccia, Antoine. *Le plan terrier de la Corse au 18ᵉ siècle: Étude d'un document géographique.* Paris: Presses Universitaires de France, 1942.

Arbellot, Guy. "La grande mutation des routes de France au milieu du XVIIIᵉ siècle." *Annales Economies, Sociétés, Civilisations* 28(1973): 765–90.

———. "Le réseau des routes de poste: Objet des premières cartes thématiques de la France moderne: Les transports de 1610 à nos jours." In *Actes du 104ᵉ Congrès des Sociétés Savantes* (1979), vol. 1. *Histoire moderne et contemporaine*, 97–115. Paris: Bibliothèque Nationale, 1980.

Armitage, Angus. *Edmond Halley.* London: Nelson, 1966.

Artz, Frederick B. *The Development of Technical Education in France, 1500–1850.* Cleveland: Society for the History of Technology, 1966.

Baker, Keith. *Condorcet: From Natural Philosophy to Social Mathematics.* Chicago: University of Chicago Press, 1975.

Berthaut, Henri Marie Auguste. *La carte de France, 1750–1898: Etude historique.* 2 vols. Paris: Service Géographique de l'Armée, 1898.

———. *Les ingénieurs géographes militaires, 1624–1831: Etude historique.* 2 vols. Paris: Service Géographique de l'Armée, 1902.

Bibliothèque Nationale. *Images de la montagne, de l'artiste à l'ordinateur: Catalogue et essais.* Paris, 1984.

Billington, David. *The Tower and the Bridge.* New York: Basic Books, 1983.

Blakemore, M. J., and J. B. Harley. "Concepts in the History of Cartography: A Review and Perspective." Monograph 26. *Cartographica* 17(1980): 1–120.

Blanchard, Anne. *Les ingénieurs du "roy" de Louis XIV à Louis XVI: Etude du corps des fortifications.* Collection du Centre d'Histoire Militaire et d'Etudes de Défense Nationale de Montpellier, no. 9. Montpellier: Université de Montpellier, 1979.

Booker, P. J. "Gaspard Monge (1746–1818) and His Effect on Engineering Drawing and Technical Education." Newcomen Society, *Transactions* 34(1961–62): 15–36.

Boud, R. C. "Samuel Hibbert and the Early Geological Mapping of the Shetland Islands." *Cartographic Journal* 14 (1977): 81–88.

————. "The Early Development of British Geological Maps." *Imago Mundi* 27(1975): 73–96.

————. "Aaron Arrowsmith's Topographical Map of Scotland and John Macculloch's Geological Survey." *Canadian Cartographer* (now *Cartographica*) 11(1974): 24–34.

Boudin, J. Ch. M. *Traité de géographie et de statistiques médicales*. 2 vols. Paris: J. B. Baillière, 1857.

Bourde, André. *Agronomie et agronomes en France au XVIII*e *siècle*. 3 vols. Paris: SEVPEN, 1967.

Boyer, Carl B. "Mathematicians of the French Revolution." *Scripta Mathematica* 25(1960): 11–31.

Bradley, Margaret. "Financial Basis of Science in Paris, 1790–1815." *Annals of Science* 36(1979): 451–91.

Brisac, Catherine. *Le Musée des Plans-Reliefs: Hôtel National des Invalides*. Paris: Pygmalion/Gérard Watelet, 1981.

Broc, Numa. "Un géographe dans son siècle: Philippe Buache (1700–1773)." *Dix-huitième Siècle* 3(1971): 222–35.

————. *La géographie des philosophes: Géographes et voyageurs au 18*e *siècle*. Paris: Orphys, 1975.

————. *Les montagnes vues par les géographes et les naturalistes de langue française au XVIII*e *siècle*. Paris: Bibliothèque Nationale, 1969.

Brown, Harcourt. "From London to Lapland and Berlin." In *Science and the Human Comedy: Natural Philosophy in French Literature from Rabelais to Maupertuis*, 167–206. Toronto and Buffalo: University of Toronto Press, 1976.

Brown, Lloyd A. "The River in the Ocean." In *Essays Honoring Lawrence C. Wroth*, 69–84. Portland, Maine: Anthoensen Press, 1951.

Buisseret, David. "Les ingénieurs du roi au temps de Henri IV." Ministère de l'Education Nationale, Comité des Travaux Historiques et Scientifiques, *Bulletin de la Section de Géographie*, 1964 (1965), 13–81.

Bunkse, Edmunds V. "Humboldt and an Aesthetic Tradition in Geography." *Geographical Review* 71(1981): 127–46.

Cailleux, André. "The Geological Map of North America (1752)." In *Two Hundred Years of Geology in America*, ed. Cecil J. Schneer, 43–52. Hanover, N.H.: University Press of New England, 1979.

Cawood, John. "Terrestrial Magnetism and the Development of International Collaboration in the Early Nineteenth Century." *Annals of Science* 34(1977): 551–87.

Charliat, P. J. "L'Académie Royale de Marine et la révolution nautique au XVIIIe siècle." *Thalès* 3(1934): 71–82.

Chartier, Roger. "Les deux Frances: Histoire d'une géographie." *Cahiers d'Histoire* 23(1978): 393–415.

Clark, G. N. *The Seventeenth Century*. Oxford: Clarendon Press, 1929; reprinted Oxford University Press, 1960.

Close, Sir Charles. *The Early Years of the Ordnance Survey*. London, 1926; reprinted, ed. John Brian Harley, New York: Augustus M. Kelley, 1969.

Clout, Hugh D., and Keith Sutton. "The 'Cadastre' as a Source for French Rural Studies." *Agricultural History* 43 (1969): 215–23.

Cohen, I. Bernard. "Jean-Joseph Delambre." *Dictionary of Scientific Biography*, 4(1971): 14–18.

Cotter, Charles H. A *History of Nautical Astronomy*. London, Sydney, and Toronto: Hollis and Carter, 1968.

Crommelin, C. A. *Physics and the Art of Instrument Making at Leyden in the Seventeenth and Eighteenth Centuries*. Leyden, 1926.

Crosland, Maurice P. " 'Nature' and Measurement in Eighteenth-Century France." *Studies on Voltaire and the Eighteenth Century* 87(1972): 277–309.

————. *Science in France in the Revolutionary Era*. Cambridge: MIT Press, 1969.

Danjon, A. "Le Verrier, créateur de la météorologie." *La météorologie*, ser. 4, no. 1(1946): 863–82.

Daumas, Maurice. *Les instruments scientifiques aux XVIIᵉ et XVIIIᵉ siècles*. Paris: Presses Universitaires de France, 1953.

Davies, Alun C. "The Life and Death of a Scientific Instrument: The Marine Chronometer, 1770–1920." *Annals of Science* 35(1978): 509–25.

Davis, John L. "Weather Forecasting and the Development of Meteorological Theory at the Paris Observatory, 1853–1878." *Annals of Science* 41(1984): 359–82.

Deacon, Margaret. *Scientists and the Sea, 1650–1900: A Study in Marine Science*. New York: Academic Press, 1971.

De Dainville, François. "Les bases d'une cartographie industrielle de l'Europe au XIXᵉ siècle." In *L'industrialisation en Europe au dix-neuvième siècle: Cartographie et typologie*, 15–33. Colloques Internationaux du CNRS, Lyon, 1970. Paris: Bibliothèque Nationale, 1972.

————. *La carte de la Guyenne par Belleyme, 1761–1840*. Bordeaux: Delmas, 1957.

————. *Cartes anciennes de l'église de France*. Paris: Vrin, 1956.

————. *Les cartes anciennes de Languedoc, XVIᵉ–XVIIIᵉ siècles*. Montpellier: Société Languedocienne de Géographie, 1961.

————. "De la profondeur à l'altitude: Des origines marines de l'expression cartographique du relief terrestre par côtes et courbes de niveau." In *Le navire et l'économie maritime du Moyen Age au dix-huitième siècle*, 195–203. Travaux du 2ᵉ colloque d'histoire maritime. Paris: Ecole Pratique des Hautes Etudes, 1958. Translated by A. H. Robinson as "From the Depths to the Heights," *Surveying and Mapping* 30(1970): 389–403.

————. *La géographie des humanistes*. Paris: Beauchesne, 1940.

————. *Le langage des géographes*. Paris: Picard, 1964.

Delaunay, Pierre. "Un projet de division géométrique du territoire français à la fin du 18ᵉ siècle." *Bulletin de la Librairie Ancienne et Moderne*, no. 121 (January 1970): 2–6.

Demangeon, Alain, and Bruno Fortier. *Les vaisseaux et les villes: L'arsenal de Cherbourg*. Brussels: Pierre Mardaga, 1978.

Destombes, Marcel. "De la chronique à l'histoire: Le globe terrestre monumental de Bergevin (1784–1795)." *Archives Internationales d'Histoire des Sciences* 27(1977): 113–34.

Devic, Jean-François Schlister. *Histoire de la vie et des travaux scientifiques et littéraires de J. D. Cassini IV*. Clermont, 1851.

Dion, Roger. *Les frontières de la France*. Paris: Hachette, 1947.

Drapeyron, Ludovic. "J. A. Rizzi-Zannoni: Son séjour en France." *Revue de Géographie* 39(1897): 401–13.

DuBus, Charles. *Démocartographie de la France des origines à nos jours*. Paris: Libraire Félix Alcan, 1931.

Ecole Polytechnique. *Livre du centenaire, 1794–1894*. 3 vols. Paris: Gauthier-Villars, 1894.

Eyles, Joan. "William Smith: Some Aspects of His Life and Work." In *Toward a History of Geology*, ed. William J. Schneer, 142–58. Cambridge: MIT Press, 1969.

Faille, René. "La carte de France divisée en carrés par Robert de Hesseln." *Bulletin de la Librairie Ancienne et Moderne*, no. 127 (August–September 1970): 122–27.

Faille, René, and Nelly Lacrocq. *Les ingénieurs géographes Claude François et Claude-Félix Masse.* La Rochelle: Rupella, 1979.

Fayet, Joseph. *La révolution française et la science, 1789–1795.* Paris: Marcel Rivière, 1960.

Ferguson, Eugene S. "The Mind's Eye: Nonverbal Thought in Technology." *Science* 197(1977): 827–36.

Foncin, Myriem. "A Manuscript Economic Map of France (End of the XVIIIth Century)." *Imago Mundi* 19(1965): 51–55.

Forbes, Eric G. "Mathematical Cosmography." In *The Ferment of Knowledge: Studies in the Historiography of Eighteenth-Century Science*, ed. Roy Porter and G. S. Rousseau, 417–48. New York: Cambridge University Press, 1980.

Fougères, M. "Les plans cadastraux de l'Ancien Régime." *Mélanges d'Histoire Sociale, Annales d'Histoire Sociale* 3(1943): 54–69.

Fuller, John G. C. M. "The Industrial Basis of Stratigraphy: John Strachey, 1671–1743, and William Smith, 1769–1839." *American Association of Petroleum Geologists, Bulletin* 53(1969): 2256–73.

Funkhauser, H. G. "Historical Development of the Graphical Representation of Statistical Data." *Osiris* 3(1937): 269–404.

Funkhauser, H. G., and H. M. Walker. "Playfair and His Charts." *Economic History* 3(1935): 103–9.

Gallois, Léon. "L'Académie des Sciences et les origines de la carte de Cassini." *Annales de Géographie* 18(1909): 194–204, 289–310.

Gilbert, E. W. "Pioneer Maps of Health and Disease in England." *Geographical Journal* 124(1958): 172–83.

Gillispie, Charles Coulston. "Laplace." *Dictionary of Scientific Biography*, 15(1978): 273–403.

————. *Science and Polity in France at the End of the Old Regime.* Princeton: Princeton University Press, 1980.

Gillmor, C. Stewart. *Coulomb and the Evolution of Physics and Engineering in Eighteenth-Century France.* Princeton: Princeton University Press, 1971.

Gingerich, Owen. "Pierre-François André Méchain." *Dictionary of Scientific Biography*, 9(1974): 250–52.

Girard d'Albissin, Nelly. *Genèse de la frontière franco-belge: Les variations des limites septentrionales de la France de 1659 à 1789.* Bibliothèque de la Société d'Histoire du Droit du Pays Flammands, Picards et Wallons, no. 26. Paris: Picard, 1970.

Guerlac, Henry. "Some Aspects of Science during the French Revolution." *Scientific Monthly* 80(1955): 93–101.

Guichonnet, Paul. "Le cadastre savoyard de 1738 et son utilisation pour les recherches d'histoire et de géographie sociales." *Revue de Géographie Alpine* 43(1955): 255–98.

Haasbroek, N. D. *Gemma Frisius, Tycho Brache and Snellius and Their Triangulations.* Delft: Netherlands Geodetic Commission, 1968.

Hahn, Roger. *The Anatomy of a Scientific Institution: The Paris Academy of Science, 1666–1803.* Berkeley and Los Angeles: University of California Press, 1971.

Hartshorne, Richard. "The Concept of Geography as a Science of Space, from Kant and Humboldt to Hettner." *Association of American Geographers, Annals* 48 (1958): 97–108.

Harvey, P. D. A. *History of Topographical Maps: Symbols, Pictures and Surveys.* London: Thames and Hudson, 1980.

Herbin, R., and A. Pebereau. *Le cadastre français.* Paris: Francis Lefebre, 1953.

Hervé, Roger. "Les plans de forêts de la grande réformation colbertienne, 1661–1690." *Ministère de l'Education Nationale, Comité des Travaux Historiques et Scientifiques, Bulletin de la Section de Géographie* 73(1960): 143–71.

Hindle, Brooke. *Emulation and Invention.* New York: New York University Press, 1981.

Howse, Derek. *Greenwich Mean Time and the Discovery of Longitude*. Oxford University Press, 1980.

Huguenin, Marcel. "Le carte des frontières de l'Est de Fort l'Ecluse à Landau." *Bulletin d'Information de l'Association des Ingénieurs Géographes*, no. 14 (July 1959): 127–36; no. 16 (March 1960): 69–89.

———. "La cartographie ancienne de la Corse." *Bulletin d'Information de l'Association des Ingénieurs Géographes*, no. 23 (July 1962): 85–98; no. 26 (July 1963): 33–55.

———. "La cartographie des Alpes françaises avant Cassini." *Bulletin d'Information de l'Association des Ingénieurs Géographes*. no. 10 (March 1958): 89–106; no. 12 (November 1958): 105–26.

———. "French Cartography of Corsica. *Imago Mundi* 24(1970): 123–37.

Jarcho, Saul. "Yellow Fever, Cholera and the Beginnings of Medical Cartography." *Journal of the History of Medicine and Allied Sciences* 25 (1970): 131–42.

Jones, Yolande. "Aspects of Relief Portrayal on 19th Century British Military Maps." *Cartographic Journal* 11(1974): 19–32.

Jouanne, R. *Les origines du cadastre ornais*. Alençon: Imprimerie Alençonnaise, 1933.

Keuning, J. "Sixteenth Century Cartography in the Netherlands." *Imago Mundi* 9(1952): 35–65.

King, Henry C., with John R. Millburn. *Geared to the Stars: The Evolution of Planetariums, Orreries and Astronomical Clocks*. Toronto and Buffalo: University of Toronto Press, 1978.

Kirchner, Walter. "Mind, Mountain and History." *Journal of the History of Ideas* 11(1950): 412–47.

Kish, George. "Early Thematic Mapping: The Work of Philippe Buache." *Imago Mundi* 28(1976): 129–36.

Konvitz, Josef W. "Alexander Dalrymple's Wind Scale for Mariners." *Mariner's Mirror* 69(1983): 91–93.

———. "Changing Concepts of the Sea, 1550–1950." *Terrae Incognitae* 11(1979): 1–17.

———. *Cities and the Sea: Port City Planning in Early Modern Europe*. Baltimore: Johns Hopkins University Press, 1978.

———. "The National Map Survey in Eighteenth-Century France." *Government Publications Review* 10(1983): 395–403.

———. "Recreating France". FMR. Forthcoming.

———. "Redating and Rethinking the Cassini Geodetic Surveys of France, 1730–1750." *Cartographica* 18(1982): 1–15.

———. *The Urban Millennium: The City-Building Process from the Early Middle Ages to the Present*. Carbondale: Southern Illinois University Press, 1985.

Laissus, Joseph. "Un astronome français en Espagne: Pierre-François-André Méchain." In 94th Congrès National des Sociétés Savantes, Paris, 1969, *Comptes-rendus, Sciences*, 1:37–59. Paris: Bibliothèque Nationale, 1970.

Landes, David S. *Revolution in Time: Clocks and the Making of the Modern World*. Cambridge: Belknap Press, Harvard University Press, 1983.

Langin, Janis. "Sur la première organisation de l'Ecole Polytechnique, texte de l'arrêté du 6 frimaire an III." *Revue de l'Histoire des Sciences* 33 (1980): 289–313.

Laussedat, Aimé. *La délimitation de la frontière franco-allemande*. Paris: Ch. Delagrave, 1901.

Lemoine-Isabeau, C. "Lettres d'ingénieurs géographes français en Flandres, en 1746." *Revue Belge d'Histoire Militaire* 13(1980): 413–26.

Lepetit, Bernard. *Chemins de terre et voies d'eau: Réseaux de transports et organisation de l'espace en France, 1740–1840*. Paris: Ecole des Hautes Etudes en Sciences Sociales, 1984.

Liddell-Hart, Basil. *The Ghost of Napoleon*. New Haven: Yale University Press, 1935.

Longhena, Maris. *L'opera cartographica di L. F. Marsili*. Pubblicazioni dell'Instituto de Geografia, ser. A, no. 3. Rome: University of Rome, 1933.

McKeon, Robert. "Gaspard-François de Prony." *Dictionary of Scientific Biography*, 11(1975): 163–66.

Mage, Georges. *La division de la France en départements*. Toulouse: Imprimerie Saint-Michel, 1924.

Marshall, Douglas W. "Instructions for a Military Survey in 1779." *Cartographica* 18(1981): 1–12.

Mellersch, H. E. L. *Fitzroy of the Beagle*. London: Rupert Hart-Davis, 1968.

Melvin, Frank E. *Napoleon's Navigational System: A Study of Trade Control during the Continental Blockade*. New York: Appleton, 1919.

Meyer, Jean, ed. *Médecins, climat et épidémies à la fin du XVIIIᵉ siècle*. Paris and The Hague: Mouton, 1972.

Morgan, Victor. "The Cartographic Image of 'the Country' in Early Modern England." *Transactions of the Royal Historical Society*, ser. 5, 29(1979): 129–54.

Musée Savoisien. *Le cadastre sarde de 1730 en Savoie*. Chambéry, 1980.

Nef, John U. *Cultural Foundations of Industrial Civilization*. Cambridge: Cambridge University Press, 1958.

Nicolson, Marjorie. *Mountain Gloom and Mountain Glory: The Development of the Aesthetics of the Infinite*. Ithaca, N.Y.: Cornell University Press, 1959.

Olmsted, J. W. "The Scientific Expedition of Jean Richer to Cayenne (1672–1673)" *Isis* 34 (1942): 117–28.

Pastoureau, Mireille. *Les atlas français XVIᵉ–XVIIᵉ siècles: Répertoire bibliographique et étude*. Paris: Bibliothèque Nationale, 1984.

———. "Les atlas imprimés en France avant 1700." *Imago Mundi* 32 (1980): 45–72.

Pedley, Mary Sponberg. "The Map Trade in Paris, 1650–1825." *Imago Mundi* 33(1981): 33–45.

———. "The Subscription List of the 1757 'Atlas Universel': A Study in Cartographic Dissemination." *Imago Mundi* 31 (1979): 66–77.

Pelletier, Monique. "La Martinique et la Guadeloupe au lendemain du Traité de Paris (10 fevrier 1763): L'oeuvre des ingénieurs géographes." *Chronique d'Histoire Maritime*, no. 9 (1984): 22–30.

Pérez-Gomez, Alberto. *Architecture and the Crisis of Modern Science*. Cambridge: MIT Press, 1983.

Perrot, Jean-Claude. *L'age d'or de la statistique régionale française (an IV–1804)*. Paris: Société des Etudes Robespierristes, 1977.

Petot, Jean. *Histoire de l'administration des Ponts et Chaussées (1599–1815)*. Paris: Marcel Rivière, 1958.

Pigonneau, Henri, and Alfred de Foville. *L'administration de l'agriculture au contrôle-générale des finances (1785–7): Procès-verbaux et rapports*. Paris: Guillaumin, 1882.

Pinon, Pierre, and Annie Kriegel. "L'achèvement des canaux sous la Restauration et la Monarchie de Juillet." *Annales des Ponts et Chaussées*, n.s., no. 19 (1981): 72–83.

Pogo, A. "Gemma Frisius: His Method of Determining Difference of Longitudes by Transporting Timepieces (1530) and His Treatise on Triangulation (1533)." *Isis* 22(1935): 469–85.

Poindron, Paul. "Les cartes géographiques du Ministère des Affaires Etrangères (1780–1789): Jean-Denis Barbié du Bocage et la collection d'Anville." *Sources, Etudes, Recherches, Informations des Bibliothèques Nationales de France*, ser. 1, no. 1 (1943): 46–72.

Porter, Roy. *The Making of Geology: Earth Science in Britain, 1660–1815*. Cambridge: Cambridge University Press, 1978.

———. "The Terraqueous Globe." In *The Ferment of Knowledge*, ed. R. Porter and G. S. Rousseau, 285–324. Cambridge: Cambridge University Press, 1980.

Pounds, Norman J. G. "The Origin of the Idea of Natural Frontiers in France." *Annals of the Association of American Geographers* 41(1951): 146–57.

Rapoport, Rhoda. "The Early Disputes between Lavoisier and Monnet, 1777–1781." *British Journal for the History of Science* 4 (1969): 233–44.

———. "The Geological Atlas of Guettard, Lavoisier and Monnet: Conflicting Views of the Nature of Geology." In *Toward a History of Geology*, ed. Cecil J. Schneer, 272–87. Cambridge: MIT Press, 1969.

———. "Government Patronage in Science in Eighteenth-Century France." *History of Science* 8(1961): 119–36.

———. "Lavoisier's Geological Activities, 1763–1792." *Isis* 58(1967): 375–84.

———. "Problems and Sources in the History of Geology, 1794–1810." *History of Science* 3(1964): 60–77.

Richardson, Philip L. "Benjamin Franklin and Timothy Folger's First Printed Chart of the Gulf Stream." *Science* 207 (8 February 1980): 643–45.

Richeson, A. W. *English Land Measuring to 1800: Instruments and Practices*. Cambridge: MIT Press, 1966.

Robinson, A. H. W. *Marine Cartography in Britain: A History of the Sea Chart to 1855*. Leicester: Leicester University Press, 1962.

———. "The Charting of the Scottish Coasts." *Scottish Geographical Magazine* 74(1958): 116–26.

———. "Marine Surveying in Britain during the Seventeenth and Eighteenth Centuries." *Geographical Journal* 123(1957): 449–56.

Robinson, Arthur H. *Early Thematic Mapping in the History of Cartography*. Chicago: University of Chicago Press, 1982.

———. "The 1837 Maps of Henry Drury Harness." *Geographical Journal* 121(1955): 440–50.

———. "The Genealogy of the Isopleth." *Cartographic Journal* 8(1971): 49–53; reprinted in *Surveying and Mapping* 32(1972): 331–38.

———. "The Thematic Maps of Charles Joseph Minard." *Imago Mundi* 21(1967): 95–108.

Robinson, Arthur H., and Barbara Bartz Petchenik. *The Nature of Maps: Essays toward Understanding Maps and Mapping*. Chicago: University of Chicago Press, 1976.

Robinson, Arthur H., and Helen Wallis. "Humboldt's Map of Isothermal Lines: A Milestone in Thematic Cartography." *Cartographic Journal* 4(1967): 19–23.

Ronan, Colin A. *Edmond Halley: Genius in Eclipse*. New York: Doubleday, 1969.

Rothrock, George A. "Musée des Plans-Reliefs." *French Historical Studies* 6(1969): 253–56.

Rudwick, Martin J. S. "The Emergence of a Visual Language for Geological Science, 1760–1840." *History of Science* 14 (1976): 149–95.

Russo, F. "L'hydrographie en France aux XVIIᵉ et XVIIIᵉ siècles." In *Enseignement et diffusion des sciences en France au dix-huitième siècle*, ed. René Taton, 419–40. Paris: Hermann, 1964.

Samoyault, Jean-Pierre. *Les Bureaux du Secrétariat d'Etat des Affaires Etrangères sous Louis XIV*. Bibliothèque de la Revue d'Histoire Diplomatique, no. 3. Paris: A. Pedone, 1971.

Sandler, C. *Die Reformation der Kartografie um 1700*. Munich: Oldenburg, 1905.

Sarton, George. "Preface to Volume XXIII of Isis (Quetelet)." *Isis* 23 (1935): 5–24.

Shinn, Terry. *Savoir scientifique et pouvoir social: L'Ecole Polytechnique, 1794–1914*. Paris: Presses de la Fondation Nationale des Sciences Politiques, 1980.

Skelton, R. A. "Captain James Cook as a Hydrographer." *Mariner's Mirror* 40 (1952): 92–119.

———. "Cartography (ca. 1750–ca. 1850)." In A History of Technology, ed. Charles Singer, 4:596–628. Oxford: Oxford University Press, 1958.

———. Maps: A Historical Survey of Their Study and Collecting. Chicago: University of Chicago Press, 1972.

———. "The Origins of the Ordnance Survey of Great Britain." Geographical Journal 128 (1962): 415–30.

Stevenson, Lloyd G. "Putting Disease on the Map: The Early Use of Spot Maps in the Study of Yellow Fever." Journal of the History of Medicine and Allied Sciences 20 (1965): 227–61.

Stilgoe, John R. "Jack o' Lanterns to Surveyors: The Secularization of Landscape Boundaries." Environmental Review 1(1976): 14–31.

Strasser, Georg. "The Toise, the Yard and the Metre—the Struggle for a Universal Unit of Length." Surveying and Mapping 35 (1975): 25–46.

Taton, Juliette, and René Taton. "Jean Picard." Dictionary of Scientific Biography, 10(1974): 595–97.

Taton, René. "César François Cassini de Thury (Cassini III)." Dictionary of Scientific Biography, 3(1971): 107–9.

———. "Gian Domenico (Jean-Dominique) Cassini." Dictionary of Scientific Biography, 3(1971): 100–104.

———. "Jacques Cassini (Cassini II)." Dictionary of Scientific Biography, 3(1971): 104–6.

———. "Jean-Dominique Cassini (Cassini IV)." Dictionary of Scientific Biography, 3(1971): 106–7.

Taylor, E. G. R. "The Earliest Account of Triangulation." Scottish Geographical Magazine 43(1942): 341–45.

———. The Haven-Finding Art: A History of Navigation from Odysseus to Captain Cook. London: Hollis and Carter, 1965.

———. "The Measure of the Degree, 300 B.C.–A.D. 1700." Geography 34(1949): 121–31.

Taylor, Kenneth L. "Nicolas Demarest and Geology in the Eighteenth Century." In Toward a History of Geology, ed. Cecil J. Schneer, 339–56. Cambridge: MIT Press, 1969.

Thompson, F. M. L. Chartered Surveyors: The Growth of a Profession. London: Routledge and Kegan Paul, 1968.

Thrower, Norman J. W. "Edmond Halley and Thematic Geo-Cartography." In The Compleat Plattmaker: Essays on Chart, Map and Globe Making in England in the Seventeenth and Eighteenth Centuries, 195–228. Berkeley and Los Angeles: University of California Press, 1978.

Trystram, Florence. Le procès des étoiles. Paris: Seghers, 1979.

Vignon, Eugène-Jean-Marie. Etudes historiques sur l'administration des voies publiques en France aux XVIIᵉ et XVIIIᵉ siècles. 4 vols. Paris: Dunod, 1862.

Vorsey, Louis de. "The Gulf Stream on Eighteenth-Century Maps and Charts." Map Collector, no. 15 (June 1981): 2–10.

———. "Pioneer Charting of the Gulf Stream: The Contributions of Benjamin Franklin and William Gerard De Brahm." Imago Mundi 28 (1976): 105–20.

Wallis, Helen. "Maps as a Medium of Scientific Communication." In Studia z Dziejow Geografii i Karbografii: Etudes d'histoire de la géographie et de la cartographie, ed. Josef Babicz, 251–62. Warsaw: Polska Akademia Nauk Zakkad Historii Nauki i Techniki, 1973.

Warner, A. J. The Sky Explored: Celestial Cartography, 1500–1800. New York: Alan R. Liss, 1979.

Waters, D. W. The Art of Navigation in England in Elizabethan and Early Stuart Times. New Haven: Yale University Press, 1958.

Whitnah, Donald R. A *History of the United States Weather Bureau*. Urbana: University of Illinois Press, 1961.

Wilford, John Noble. "Prints of Franklin's Chart of Gulf Stream Found." *New York Times*, 6 February 1980, A1, B7.

Williams, L. Pearce. "Science, Education and the French Revolution." *Isis* 44 (1953): 311–30.

Wolter, John. "The Heights of Mountains and the Lengths of Rivers." Library of Congress, *Quarterly Journal* 29 (1972); reprinted in *Surveying and Mapping* 32(1972): 313–29.

Woodward, David. "The Study of the History of Cartography: A Suggested Framework." *American Cartographer* 1 (1974): 101–15.

Woolf, Harry. *The Transits of Venus*. Princeton: Princeton University Press, 1959.

Index

Centralization in France (*continued*)
130–31, 145, 152, 158–59; and transportation
maps, 105, 112–15, 120–22
Chabert, Joseph-Bernard, 42, 73
Chanlaire, P. G., 99
Chappe d'Auteroche, Jean, 74
Cherbourg, harbor mapped, 98
Chevalier, François, 19
Cheysson, J. J. Emile, 158
Chézy, Antoine de, 108, 136, 141
Claret de Fleurieu, Charles Pierre, 74–76
Colbert, Jean-Baptiste, 1, 4, 6
Collections of maps: of the crown, 165 n.7; by
d'Anville, 33–35; of the Ponts et Chaussées,
113; during the Revolution, 54–56
Condorcet, Marie Jean, Marquis de, 42, 45, 108,
116, 122
Contour lines: showing depths in water, 67, 71,
75, 77–81; showing land elevation, 91, 96–101,
116, 119, 134–35
Cook, James, 72, 74
Coquebert de Montbret, Charles Etienne, 131–34
Cornuau, Pierre, 41, 116
Corsica, 42, 90
Coulomb, Charles Augustin de, 108
Coutans, Dom Guillaume, 113
Croix, Chrestien de la, 35–37, 99
Crome, A. F. W., 129
Cruquius, Nicolas, 68
Cuvier, Georges, 89

Dalrymple, Alexander, 26, 35, 71–72, 76
d'Angeville, Adolphe, 149–50
d'Anville, Jean-Baptiste Bourguignon, 19, 33–35
d'Après de Manevillette, J. B., 73
d'Arçon, Jean-Claude Le Michaud, 39–40, pl. 3
De Brahm, William Gerard, 64
De Kermadec de Moustier, M., 106–8
Delagrive, Abbé Jean, 109
Delambre, Jean-Baptiste, 26, 46, 50–51, 57, 167 n.47
Delisle, Guillaume, 70
Delisle, Jean-Nicolas, 74
Demarest, Nicolas, 56, 87–89
Digges, Leonard, 3
Dodson, James, 71
Donnant, Denis-François, 130
Du Carla, Marcellin, 77–82, 98, 101, 153
Dufrenoy, O. P., 89
Duhamel du Montceau, Henri-Louis, 78
du Muy, Louis, 97
Dupain-Triel, Jean-Nicolas, Jr., 79, 100, 119–20
Dupain-Triel, Jean-Nicolas, Sr., 77, 87, 90, 119
Dupin, Charles, 146, 149
Dupuis-Torcy, Pierre-Louis, 122
Dutillet de Villars, M., 42

Ecole de Géodésie Pratique, 50
Ecole des Ingénieurs Géographes, 140
Ecole des Ponts et Chaussées, 47, 135–39, 141–42,
152, 154, 156
Ecole Polytechnique, 138–42, 154
Education, cartography in: for army engineers, 39,
139–40; for civil engineers, 135–42, 152, 156; for
hydrographers, 73; for nonprofessionals, 56
Elie de Beaumont, Jean, 89
Engineering: contrasted with science, 82–83, 85,
98, 101; and thematic cartography, 135, 142,
152–58. *See also* Education
England, compared with France in cartography,
1–3, 26–28, 46, 71–73, 122–23, 143, 145, 151
Espey, James P., 144

Fabre, Jean-Antoine, 116
Fer de la Nouerre, Nicolas, 118–19
Financing cartography: cadaster, 42–43, 52, 58–61;
d'Anville collection, 34–36; geological maps, 87;
national map survey, 22, 24–25, 30–31;
oceanographic charts, 77
Fitzroy, Robert, 145
Folger, Timothy, 63–64
Foreign affairs, and border mapping, 32–38
Forfait, Pierre, 110
Fourcroy, Antoine-François de, 108
France, conditions affecting cartography: and
Cassini survey, 9, 18; during Restoration, 61–62,
122–23, 142, 158–59; on eve of Revolution, 29–
31, 40–41; during Revolution, 37–38, 43–50
Franklin, Benjamin, 63–66
Frère de Montizon, A., 146
Frisius, Gemma, 2
Frissard, Pierre-François, 131

Galton, Francis, 145
Gascoigne, William, 3
Gauthey, E.-M., 118
Geodesy, 2, 10–13, 26–28, 46–51, 58–61, 76, 84, 86.
See also National map survey
Geological maps, 82–91, 131–33
Girard, Pierre-Simon, 100
Girard-Soulavie, Jean-Louis, 90
Goyon de la Plombaine, Henri de, 115
Grandjean, Jean-Sebastien, 35–37
Grandjean de Fouchy, Jean-Paul, 78
Greenough, George Bellas, 89
Greenwich Observatory, 27–28
Guerry, André-Michel, 148–49
Guettard, Jean-Etienne, 84
Gulf Stream, 63–66
Gunter, Edmund, 3

Halley, Edmond, 3, 64, 70
Happel, Eberhard Werner, 63